Waste

Resources series

Gavin Bridge & Philippe Le Billon, *Oil*, 2nd edition
Anthony Burke, *Uranium*
Jennifer Clapp, *Food*, 2nd edition
Peter Dauvergne & Jane Lister, *Timber*
Elizabeth R. DeSombre & J. Samuel Barkin, *Fish*
Kate Ervine, *Carbon*
David Lewis Feldman, *Water*
Gavin Fridell, *Coffee*
Derek Hall, *Land*
Andrew Herod, *Labor*
Kristy Leissle, *Cocoa*
Michael Nest, *Coltan*
Kate O'Neill, *Waste*
Bronwyn Parry and Beth Greenhough, *Bioinformation*
Ben Richardson, *Sugar*
Ian Smillie, *Diamonds*
Adam Sneyd, *Cotton*
Mark Thurber, *Coal*
Bill Winders, *Grains*

Waste

KATE O'NEILL

polity

First published in 2019 by Polity Press
Reprinted: 2019 (twice)

Polity Press
65 Bridge Street
Cambridge CB2 1UR, UK

Polity Press
101 Station Landing
Suite 300
Medford, MA 02155, USA

ISBN-13: 978-0-7456-8739-1
ISBN-13: 978-0-7456-8740-7(pb)

A catalogue record for this book is available from the British Library.

Library of Congress Cataloging-in-Publication Data
Names: O'Neill, Kate, 1968- author.
Title: Waste / Kate O'Neill.
Description: Cambridge, UK ; Medford, MA, USA : Polity Press, 2019. | Includes bibliographical references and index.
Identifiers: LCCN 2018061778 (print) | LCCN 2019000572 (ebook) | ISBN 9780745687438 (Epub) | ISBN 9780745687391 (hardback) | ISBN 9780745687407 (pbk.)
Subjects: LCSH: Refuse and refuse disposal--Economic aspects. | Recycling (Waste, etc.) | International economic integration.
Classification: LCC HD4482 (ebook) | LCC HD4482 .O54 2019 (print) | DDC 363.72/8--dc23
LC record available at https://lccn.loc.gov/2018061778

Typeset in 10.5 on 13pt Scala by
Servis Filmsetting Ltd, Stockport, Cheshire
Printed and bound in the United States by LSC Communications

For further information on Polity, visit our website:
politybooks.com

For My Parents

Bob and Sally O'Neill

For encouraging a lifelong interest in trash and where it goes (and for childhood trips to the Canberra tip, where, according to my dad, you could rely on finding diplomats, politicians and retired admirals prospecting for secondhand treasures)

Contents

Figures, Tables, and Boxes

Abbreviations

BAN	Basel Action Network
Bo2W	Best of Two Worlds
BPA	bisphenol A
CE	circular economy
EPR	extended producer responsibility
EU	European Union
GAIA	Global Alliance for Incineration Alternatives (now known simply as GAIA)
GlobalRec	Global Alliance of Waste Pickers
GWMO	Global Waste Management Outlook
HS Code	Harmonized System
ISRI	Institute for Scrap Recycling Industries
ISWA	International Solid Waste Association
MRF	materials recovery facility
MSW	municipal solid waste
mt	million tons (metric)
NGO	non-governmental organization
NRDC	National Resources Defense Council
OECD	Organization for Economic Cooperation and Development
RIC	resin identification code
RIOS	Recycling Industry Operating Standard
SDGs	Sustainable Development Goals
StEP	Solving the E-waste Problem
SWANA	Solid Waste Association of North America
UNEP	United Nations Environment Programme

UNFAO	United Nations Food and Agriculture Organization
USEPA	US Environmental Protection Agency
WIEGO	Women in Informal Employment: Globalizing and Organizing
WRAP	Waste and Resource Action Programme
WTE	waste-to-energy
WTO	World Trade Organization
ZW	Zero Waste

Preface and Acknowledgments

Sitting in my office many nights, weekends, and holidays, and writing about waste was something of a wrenching experience. As anyone who has taken the lid off their reusable coffee mug to see the remnants of last week's latte staring right back at them knows, living a low-waste life is challenging for those of us who grew up in cultures of disposability. We make decisions about what we throw out, what we recycle, and what we buy or do not buy in the first place all the time, although we do not always think about the consequences of our decisions. This often means those discarded items are shipped to places overseas to be dismantled or recycled and resold.

This book is designed to make these journeys visible. It draws attention to the global markets that exist for discarded goods, and the livelihoods that depend on extracting value from what others have thrown away. It also makes visible the risks attendant on this new resource frontier, the growth of global activism around waste and recycling, and how actors engaged in governing these transactions are responding. It is less about individual consumer choices and behavior than about the systems in which our practices are embedded.

I am writing this at a time when levels of wastes – particularly plastic wastes – are higher than they ever have been on global, national, and local political agendas, and it is clear that a lot will change over the next few years. It is also true that I have had to leave a lot out of this short book (readers are encouraged to follow up on issues that interest them on their own). Despite all the challenges and problems, at the end of this writing process I felt moderately optimistic about our ability to address the global waste crisis, even if we can't solve it.

Despite all the hours of solitary writing, this book is really the product of many interactions and conversations. One of the things I enjoy about working on waste is that anyone I talk to, anywhere, about what I do has a story to tell or a comment to make. Many of these made it into this book or informed the directions I chose. Therefore, first of all, I would like to thank everyone who took the time to engage with me at conferences and presentations, in class and on campus, at the gym, at airports, in restaurants, on hikes, etc., over the past several years and share their experiences and opinions on the topics in this book.

For particular help with reading, editing, and commenting on draft chapters, my special thanks to Alastair Iles, Erin Bergren, Raul Pacheco-Vega, and Manisha Anantharaman. I would also like to acknowledge my undergraduate research assistants, including Leila Hooshyear, Sierra Westhem, and Aubrey Hills. Thanks to Peter Dauvergne, Shannon Davis, Emily Polsby, Freyja Knapp, Louise Mellor, Aaron King (Seven Seas Hauling), Amy Mason, Scott Silva, Anna Yip, and the UC Berkeley Zero Waste community for inspiration. Josh Lepawsky and the editors at *Welt-Sichten: Magazin für globale Entwicklung und ökumenische Zusammenarbeit* helped me with figures and data. Jenny Weeks at *The Conversation* published my earlier pieces on China and Operation National Sword, which I drew on extensively. Likewise, many great people in the waste and discard studies communities on Twitter helped me out at various times with many queries, small and large, and I am immensely grateful for this community. Thanks also to Louise Knight, Sophie Wright, and Nekane Galdos at Polity Press for being so patient with me, and to two anonymous reviewers for their helpful comments.

Finally, thank you to my husband, Wil Burns, for helping out with ideas, support and logistical advice at many times during this process. Not to mention putting up with my ongoing commentary on what we should or should not be recycling (and in what condition) at any given moment.

CHAPTER ONE

The Global Political Economy of Waste

This book is about waste – "what we do not want or fail to use" (Gourlay 1992) – as a global resource, a livelihood, and a source of risk. While other resources – timber, minerals, fish stocks – are coming under tremendous strain, wastes produced across all sectors of economic activity are growing in volume and in their potential for profit. A study published in *The Anthropocene Review* in 2017 estimated the cumulative total of the material output of collective human enterprise since the Industrial Revolution at 30 trillion tons (Zalasiewicz et al. 2017). Much of this accumulated stuff is still with us and creating a new geological stratum, of trash. This study, while speculative, makes two things clear: wastes do not disappear, and they are a potential reservoir of extractable resources.

Wastes are highly differentiated. Sewage, batteries, construction waste, discarded clothes, scrap paper and plastic, and nuclear waste all belong in this same broad category, but have very different origins, lifecycles, impacts, and values. They exist in households, in landfills, in factories, in the oceans, and in outer space. Satellite debris clogs the outer atmosphere, while old electronics pile up in peoples' attics and basements. Wastes flow into the global commons: plastics flood the oceans and landfill gas emissions exacerbate climate change. Some wastes – particularly organic wastes such as sewage – biodegrade quickly; others have a far longer half-life. Newspaper, banana peels, or orange peels can decompose in a matter of weeks under the right conditions. However, this process takes hundreds of years for some of the most common non-organic wastes, including plastics, ceramic, and glass. Nuclear waste (and some chemical wastes) stays toxic for tens of thousands of years, to the point where government authorities have considered creating symbols for "hazard" or "danger" that will be decipherable to humans once our own civilization has crumbled. One of the symbols considered by a US Department of Energy working group was a facsimile of Edvard Munch's famous painting, *The Scream*.[1]

Wastes are not just thrown away. They are reused, recycled, or reprocessed for the valuable elements they contain, and to protect our overburdened environment. Discarded electronics contain copper, gold, and other metals. Iron and steel pipes and girders can be extracted from the rubble of demolished buildings. Nutrients and energy can be obtained from discarded food. Creating a global circular economy, which in its ideal form would see nothing discarded and everything reused, could allow us to live within planetary limits.

Evaluating something as a waste or a resource, even in everyday transactions, is, however, highly subjective. It depends on context and perceptions. As an illustration, some years ago a friend of mine moved with her family to Zambia. After they had unpacked their cardboard moving boxes, someone came to their door and asked to collect them. They negotiated a price. It turned out that while my friend thought the price was what they would pay to get rid of the boxes, it was in fact what the collector wanted to pay them to take and reuse the boxes. This example, at a small scale, illustrates how one person's waste is, to another, a valuable resource or commodity, creating complications for transactions, markets, and governance. This book is, therefore, also about how complicated a resource waste is. It also poses risks to those who deal with it: to workers, local communities, those who produce it, and those who ship it. Its value varies with even slight fluctuations in market conditions. These factors create the need for governance that can take these complexities into account. Such governance is still under-developed.

The Rise of the Global Waste Economy

Economic growth and industrialization, especially in the twentieth century, transformed the relationship between wastes and resources in the industrialized world. With the expansion of the global economy after World War Two, people in wealthy nations could, for the first time, experience disposable consumption on a mass scale, generating what is now known as municipal solid waste (MSW). MSW includes plastic, paper, metal, and other non-organic wastes generated by households and businesses. Large-scale industrial production generated its own waste, including metal, concrete, and glass. In addition, the era was marked by massive increases in the production of chemical wastes, many of them toxic and highly persistent (lasting for a very long time) in the environment.

Waste became a problem to be dealt with, rather than something to be collected and reused. Landfills and waste incineration facilities expanded in size and number. Large-scale municipal and industrial waste collection, removal, and disposal services meant that for most middle-class and upper-middle-class communities, wastes were to be taken out of sight and out of mind. As a result, the problem of waste followed the same pattern as other types of environmental risk, where the costs moved away from people with socioeconomic power and towards people with less

power and money. With the rise of free trade and economic globalization came opportunities to ship wastes,
especially hazardous wastes, overseas to poorer countries
or indeed among richer ones (Vallette and Spalding 1990).
In the 1990s up to 90 percent of the hazardous waste trade
was legal and carried out between member states of the
Organization for Economic Cooperation and Development
(OECD), but still imposed risks on recipient communities
(O'Neill 2000). Resistance to waste facilities helped spark
the environmental justice movement in the US (Bullard
1991), and movements and campaigns across Europe and
around the world. International non-governmental organizations (NGOs) including Greenpeace International and
the Third World Network combatted the "toxic trade" –
waste dumping from rich to poor countries.[2]

Campaigns against the hazardous waste trade revealed
wastes' global reach, and the extent to which people in
the wealthy North were outsourcing their risk and the
costs of waste disposal. Subsequently, literal mountains of
solid waste in mega-landfills around the world, piles of old
electronics and computers and, most recently, devastating
pictures of discarded plastics clogging the world's oceans
have brought home the extent of this crisis. The impacts on
communities living on and around these sites fueled global
activism.

These revelations forced new thinking about extracting
and recycling items of value from waste streams at a far
larger scale than in previous times, from goods to metals
to energy. Informal workers and large multinational corporations look to "urban mines" to extract resources and
make a living. They also propelled thinking about how to
reduce this waste stream, reducing its flow, and diverting
its contents away from final disposal and back into productive use.

Several factors have driven this new global waste economy. First, we can now estimate how much value is trapped in waste. For instance, the total value of all raw materials present in discarded electronics was estimated at approximately €55 billion in 2016 (see chapter 4; Baldé et al. 2017). Large multinational corporations as well as local trash pickers have direct economic interests, driven by the need for recycled raw materials, new energy sources, and the prospect of extracting gold, copper, and other valuable metals, or simply for a living pulling out useful parts and objects from others' discarded goods.

Second, human production and consumption generate too much waste, to the extent that (despite our inventiveness), we are running out of space to put it. Higher levels of wealth, consumption, production, and population growth rates help explain the rise of waste generation, as do faster rates of urbanization, particularly in developing countries.

Two recent studies attempt to quantify global wastes and the challenges they pose. The *Global Waste Management Outlook* (Wilson et al. 2015), was produced by the UN Environment Programme (UNEP) and the International Solid Wastes Association (ISWA), an industry association and a leading authority in this area. It estimates total global production of municipal, commercial, and industrial wastes and waste from construction and demolition at around 7–10 billion tons per year.

What a Waste: A Global Review of Solid Waste Management in its first and second editions (Hoornweg and Bhada-Tata 2012; Kaza et al. 2018) was produced by the World Bank. It estimated that MSW production has risen ten-fold in the past century. In 2010 the world produced 3.5 million tons per day; by 2025 that total could reach six million tons per day. Rates of waste generation will rise most steeply in Africa and South Asia, overall and per capita. China produced 520,550 tons per day in 2005 but could produce 1.4 million tons daily by 2025. Under business-as-usual scenarios, we will not reach a point of global "peak waste," when urban populations stabilize and waste generation levels off, until the next century (Hoornweg et al. 2013). In the second edition of this report (Kaza et al. 2018), the authors predict that waste generation will outpace population growth by more than double by 2050 (p. xi), reaching a total of 3.4 billion tons annually.

Increased use of disposable products, such as single-serving plastic drink bottles or packaging waste, has added to the world's trash heaps. In 2017, *The Guardian* reported that a million plastic bottles are bought around the world every *minute* and barely 10 percent of those are recycled back into bottles. Such wastes are dumped into "monster" landfills in and around growing mega-cities such as Mexico City, Beijing, and Lagos. The world's 50 largest open landfills directly affect the daily lives of 64 million people who live nearby (ISWA 2016, p. 16). Millions of tons of plastics have spilled into the oceans, where they persist for hundreds of years, spinning in massive gyres. A 2018 study estimated that at least 79,000 tons of ocean plastic are floating inside an area of 1.6 million square kilometers, an area twice the size of Texas or three times the size of France, depending on one's point of view (Lebreton et al. 2018).

Third, changing pressures and patterns of globalization have expanded the reach of the global waste economy. The volume of the global waste trade has significantly increased, and the types of wastes shipped have diversified. Waste supply chains are lengthening, spanning thousands of miles across continents rather than hundreds of miles across a neighboring border.

Traditional perceptions of the waste trade as a North-to-South problem have broken down as it has become apparent that waste supply chains are now more complex than 20 years ago. These trends have been driven by open global trade rules, greater movement of goods, cheaper shipping, and the outsourcing of cheap labor needed for basic recycling and disposal processes to countries in the global South. They have also galvanized activism around the world, as waste pickers (informal waste workers) and other global activist groups mobilize against the trade. For example, the South–South trade in e-waste is overtaking the North–South trade, overturning popular perceptions of perpetrators and victims.

In 2018, a single seismic event reshaped the global politics of waste and made many aware of how much waste recycling and disposal has become a global business. Until early 2018, China took in close to half of the plastics thrown into recycling bins in the US and other wealthy nations (along with many other, higher quality types of scrap), to feed its growing manufacturing centers. In 2017, it announced it would effectively halt this practice. It had received too much plastic, paper, and other low-quality scrap, often too contaminated to easily reprocess, and was tired of being seen as the "world's garbage dump." Chapter 6 goes into this case in depth, but "Operation National Sword," as this policy is called, sent shockwaves through recycling and waste management industries, and demonstrated how vulnerable the global waste economy, the subject of this book, is.

Themes of the Book

This book has three overarching themes: the emergence of wastes as a global resource frontier, the magnified risks that attend this process, and the governance challenges (and innovations) these two together have generated. These trends are often exemplified in the global trade in different sorts of waste, and in shifting patterns of foreign direct investment in waste management and resource extraction around the world.

The Global Resource Frontier

Wastes, on a large scale, have become one of the planet's newest global resource frontiers. As early as 1969, an undersecretary of the US Department of the Interior told a waste management seminar in Houston that "trash is our only growing resource" (Crooks 1993, p. 22). In 2011, the Bureau of International Recycling, the global scrap industry association, proclaimed "the end of the waste era," a statement echoed by environmentalists, industry leaders, and politicians – that wastes are no longer "unwanted or surplus to requirements." Instead, past – and present – wastes will help fuel a richer and more sustainable future.

Demand for extracted materials to be reprocessed or recycled into new products or inputs for industrial and consumer use is rising in the world's fastest growing economies. The rise of the e-waste trade and the extent to which China became the Western world's major repository for plastic, paper, and other scrap exemplify this trend. The global frontier is occupied by workers, by municipal authorities, multinational corporations from waste, energy and mining sectors, international organizations, activist groups, and a whole network of brokers who oversee shipping of wastes from point of production to point of disposal.

Resource frontiers open at the limits of scarcity. As Edward Barbier points out, economic growth is not just about the allocation of scarce resources, it is also about finding and exploiting new resource frontiers and shipping those resources back to where they will be processed and sold. A frontier as used here and in relevant fields is "an area or source of unusually abundant natural resources and land relative to labor and capital" (Barbier 2015, p, 57; see also Peluso 2017). Sometimes the frontier is spatial: the Arctic, as sea ice melts due to climate change, is described as a new resource frontier for oil and other mineral resources. Sometimes technology opens the frontier. Deep seabed mining has only recently become a real possibility, although – with oil prices low and easier to reach sources abundant – it is not yet cost-effective to undertake.

The term frontier is not neutral. Its use deliberately harkens back to gold rushes in the mid-nineteenth century around the Pacific Rim and to the race for land by settlers in the US West, regardless of the people already living there. It evokes competition, conflict, even violence, and the displacement of existing resource users or local communities. It also evokes the possibility of large profits reaped in the absence of institutionalized governance.

Waste is a global resource frontier in singular ways. Joseph Schumpeter describes development in this context as "the conquest of a new source of supply of raw materials . . . irrespective of whether this source already exists or whether it has first to be created" (Schumpeter 1961, p. 66, quoted in Barbier). It does not have a specific geographic location, let alone a distant one. Waste is all around us, in landfills, incineration facilities, attics and basements, and open dumps in or near all the places human beings live, for decades. Wastes are also highly mobile. As subsequent case study chapters show, electronic, plastic, and food wastes and scrap (along with used clothing, second-hand cars, and used tires) are easily and often shipped across and within national borders.

This frontier has been opened by economic need (the real or anticipated scarcity of virgin resources), technology (waste-to-energy, metal extraction techniques), and by an abundance of capital and cheap labor. Competition and conflict at this frontier happen as corporations and communities compete for market access, market shares, and livelihoods. As with other natural resources, processes of enclosure, such as capping a landfill or fencing it off, exclude traditional users who may have treated that resource as common property, and as their livelihood.

Waste's exploitation has an environmental as well as an economic rationale. Corporate, government, and civil society actors frame waste exploitation and reuse as critical components of a sustainable, or green, economy. Waste-to-energy companies highlight their role in reducing greenhouse gas emissions from traditional energy sources. Multinational corporations engaged in urban mining and small-scale municipal recycling companies alike tout the environmental services they provide through extracting or reusing metals.

The nature of wastes as a global resource frontier found in and around human settlements highlights how local communities, authorities, waste workers, and other local actors are embedded in the global waste economy. Waste is not a unique resource in that respect, but its local ubiquity and its ability to travel the globe highlight its special characteristics. Waste needs to be re-imagined as a global resource, not a local problem. Making this frontier visible is the only way to create effective governance mechanisms that enable the reuse of these valuable resources while mitigating the magnified risks outlined in the following section. We see it in the cases addressed in this book, as we look at waste work and labor, particularly in the global

South, and at the international trade in electronic wastes and in scrap, from steel to plastic and paper.

Magnified Risks

Second, wastes have always been characterized by risk, even as they are treated more and more as a resource. These risks have become magnified as waste generation, movements, and impacts have grown – and globalized – over the past decades. The negative impacts of wastes have always disproportionately affected the most economically disadvantaged, who often belong to racial or ethnic minority groups, but geographic distancing of waste disposal from its point of production has exacerbated these risks.

"Distancing" is literal: waste can end up, through trade or the circulation of ocean currents, thousands of miles from where it is generated. It is also figurative, disconnected from and out of the thoughts of those who produce the waste in the first place. Few question where their waste plastic, paper, and electronics go, or who takes care of it. Distancing as a concept helps make sense of the inequities of our globalized world (Clapp 2002; Princen 2002). It applies equally well to global consumption and production, and the supply chains that link them.

Risks associated with waste management and the *solutions* to the waste crisis have magnified too. Mega-landfills sprawl around the world's largest and fastest growing cities. They are massive and support large communities. For example, Jam Chakro in Pakistan covers 202 hectares (500 acres). It supports an informal recycling community of 5000 and five million people live within a ten-kilometer radius. Bantar Gebang in Indonesia takes 230,000 tons of waste per year, and already holds 28–40 million tons. These are only two of the world's 50 largest "monster-dumps" (ISWA 2016).

Massive accumulations of wastes pose very literal risks. People have been killed in "waste slides" around mega-dump sites. From December 2015 to June 2016, over 750 deaths worldwide could be directly attributed to poor waste management in dumpsites (ISWA 2016). Smoke from burning waste causes further health hazards. Fires in tire dumps, for example, can blaze for weeks, months, even years, discharging toxins and oil residue into the air and earth around them. The smoke from a 2012 fire in a dump in Kuwait that contained over five million tires could be seen from space.

Recycling plants have moved from the US and Europe to sites with cheaper labor, and laxer regulations, in South and East Asia, and in Africa. Waste work remains difficult and dangerous, and the handling, processing, and recycling of municipal and industrial wastes is often carried out by informal sector workers, or waste pickers, unprotected by health and safety regulations or job security. Waste crime extends across national borders: Interpol, the international policing agency, tackles e-wastes illegally trafficked across borders as part of its environmental crimes division. Incineration too, even when touted as clean, climate-friendly waste-to-energy production, can harm local communities if emissions are not controlled.

The most hazardous wastes persist, leaving a toxic legacy for generations to come. Not a single gram of high-level nuclear wastes has been safely disposed of or sequestered for the long term. Spent nuclear fuel rods continue to be stored in pools around nuclear power plants. The 2011 Fukushima Daiichi nuclear power plant disaster is a case in point. A series of meltdowns were triggered after the Tohoku earthquake (9.0 on the Richter scale) and tsunami shut down the emergency power generators. When spent fuel rods in cooling ponds next to the plant overheated, significant quantities of radiation were released into the atmosphere. Over 150,000 nearby residents were evacuated (most of whom have never returned), radiation was found in food produced over 300 km away, and the accident significantly diminished the prospects for new nuclear power plants worldwide.

Increased generation of MSW in urban areas, particularly in developing and emerging economies, strains existing collection infrastructure. At least 2 billion people worldwide still lack access to solid waste collection and at least 3 billion people lack access to controlled waste disposal facilities (Wilson et al. 2015). In 2015, Lebanese authorities sparked a crisis when they closed a major landfill. Pictures of a river of trash in the Beirut suburbs – a waterway full of white plastic garbage bags – flashed around the world in 2015, although they could not convey the stench that permeated the city or the threats to public health the trash posed. A contract between the government and a British company, which would have shipped the waste to Russia, had fallen through, leaving an ongoing crisis, and conflict between Lebanon's government and its population.

Finally, global markets for secondary materials are fragile. They remain subject to the ups and downs of the global economy and the prices of virgin material, or to sudden changes in the policies of a dominant buyer like China (Jolly 2007; Xiarchos and Fletcher 2009). For one, scrap metals have not, unlike primary metals, historically had futures markets, which serve to stabilize prices in the face of shocks and short-term volatility. Scrap metals have smaller margins of profitability than primary metals, therefore price changes affect their markets far more.

This vulnerability affects both large scrap brokers in the US and informal workers in South Africa who must walk miles to sell scrap metal they collect (Schenck et al. 2017).

The 2008 market crash left thousands of tons of scrap metal stranded in ports. China's enforcement of restrictions on scrap imports left queues of container ships loaded with discarded plastics in limbo, waiting outside ports. In a globalized market, exogenous and endogenous shocks have rapid knock-on effects on workers, industry, and local governments, who lack the capacity to shore themselves up against these risks.

Governance Challenges and Innovations
The global waste crisis is highly complex. It extends across international borders and into the oceans, the atmosphere, and even outer space, but it is also hyper-local as wastes pile up on streets and communities mobilize against incinerators and landfills in their neighborhoods. Many actors are involved in the business of producing, collecting, disposing of, and recycling wastes, often with directly conflicting interests. Accurate data are hard to obtain, making responsibility for producing waste-related pollution hard to pin down. Wastes have built up around the world to an extent beyond our capacity to deal with them. They are piling up in the growing mega-cities in Asia and Africa that do not have the infrastructure to deal with them. They inflict environmental injustices and, as we shall see in later chapters, this leads to contention about how to deal with them, whether as a risk (waste) or a resource (scrap).

Understanding waste as a globalized resource frontier has fundamentally challenged traditional paradigms of waste governance. It can no longer be governed simply as something at the end of its useful life. But can wastes be grouped with governance of traditional extractive resources, such as timber, oil, or minerals? This understanding of waste also makes the global politics of waste highly visible. Technology will not solve our waste problems if the worst risks can more cheaply or easily be foisted on others living thousands of miles away. People engaged in global waste governance now need to figure out how to extract and utilize resources contained in wastes while minimizing risks to the vulnerable and sharing benefits.

The chapters that follow examine these governance challenges at different jurisdictional levels, and how they are starting to be met. Campaigns against food waste have led to innovative governance solutions. Various actors, including the scrap industry are considering enabling labeling, or certification, of scrap safe to ship across borders. Cities are taking the lead in implementing Zero Waste initiatives, and activist alliances are transforming the role of informal sector workers at home and globally. Existing international agreements, such as the 1989 Basel Convention on the Control of Transboundary Movements of Hazardous Wastes and Their Disposal, are proving inadequate to govern the new global waste economy. The actors developing and implementing the 2015 UN Sustainable Development Goals (SDGs) have embraced waste reduction and zero waste targets, but the SDGs are non-binding normative goals, not hard directives that nation states must comply with.

The waste activist landscape has also transformed. Activist groups influence governance initiatives and create their own. New networks span the gap between global NGOs and community-based groups. These include the Global Alliance of Waste Pickers (GlobalRec), and GAIA, which campaigns against incineration and for zero waste alternatives. The Basel Action Network (BAN), which has long opposed the hazardous waste trade and advocated for a ban, now focuses on ending e-waste trading from North to South. A growing global "right to repair" movement is pushing against planned obsolescence in the electronics industry. As the crisis of plastic pollution in the oceans unfolds, ocean-related organizations have teamed up with waste and chemicals groups to push the United Nations to take action. Major NGOs – such as the Natural Resources Defense Council and the World Resources Institute – have added food waste to their portfolio. Research NGOs, such as UK-based Waste and Resource Action Programme (WRAP), do invaluable work in increasing knowledge and political transparency around consumer waste.

Zero Waste activism has also achieved critical mass in the past decade or so, visible on college campuses, in municipal policy programs, and at high level discussions of sustainable development goals in the United Nations. Zero Waste activists push for policies and programs aimed at maximizing diversion of wastes away from landfills towards recycling and reuse, and reduction of waste pro-duction. High-profile foundations, most notably the Ellen MacArthur Foundation (founded in 2010) have taken on building a global circular economy as their mission. Entities such as the Closed Loop Fund and Foundation leverage private and public funding to build new waste infrastructure in the US and overseas. Waste associations, such as ISWA (the International Solid Waste Association), SWANA (the Solid Waste Association of North America) and ISRI (the Institute of Scrap Recycling Industries) rep-resent their industries, do data gathering and research, and push for sustainable waste management solutions.

Existing waste governance measures, from the global community or national and local governments and com-munities, are often controversial. Crafting governance initiatives within the new global waste economy must take into account many competing interests and influences. Waste trading (of all sorts) is a particularly thorny exam-ple. Banning shipment of wastes across national borders, especially from richer to poorer countries, has long been the mainstay of global governance efforts, but it has failed, except in the worst cases of hazardous waste dumping. If there is one lesson that can be learned from looking at wastes as a globalized resource, it is that they go where the markets are – where they can be bought, sold, and used. These manufacturing bases are rarely in the global North. A global recycling economy will depend on waste or scrap moving in response to demand, not to supply. This means somehow the international community needs to find a way to make a waste trade work safely and effectively, or a way to shut down waste production at source. Finding the right site for global governance of waste, whether it is the Basel Convention or the World Trade Organization (see chapter 6) is an ongoing challenge.

Perspectives on Wastes

To more fully comprehend the global waste and recycling economy, we need to deploy a range of perspectives on wastes, above and beyond risks and resources. These include wastes as externalities, commodities, livelihoods, and even as inputs for production. We apply all these perspectives in subsequent chapters to help make sense of the complex economies of waste work, food, e-wastes, and plastic scrap. This typology draws on the conceptual framework on the role of wastes in society developed by geographer Sarah Moore (Moore 2012).

Historically, we have viewed wastes as *externalities*, or unintended byproducts, of production, whose collection, management, and disposal is designed to be an invisible service, in practice (wastes are disposed of permanently and out of sight), and politically (see chapter 2). Waste producers pay to have the waste hauled away. As a negative externality, waste imposes higher costs on local communities and the environment than it does on the waste producers themselves, which gives them little incentive to pay to deal with it at source or reduce generation. Indeed, many of those responsible for production, transport, and disposal hope that their actions will not come under scrutiny. This perspective has dominated waste management systems, technology – landfills, incinerators – and governance for decades.

Wastes can also pose significant *risks* or hazards. Growth of chemicals production and use introduced toxins and pollutants into widely used products. These risks are exacerbated if the waste is not managed safely. Often the hazards are concentrated in and around poor, ethnic, or racial minority or otherwise marginalized communities. Recently, waste generation and storage has been shown to contribute to greenhouse gas emissions, hence waste management, use, and reduction is now a weapon against climate change.

Third, waste is an important *resource*. Wastes contain valuable resources, available for use and extraction. Waste is a type of ore, containing valuable metals to be extracted, resold, and reused. It can be a source of energy, like oil or gas. It can be refurbished and reused in construction. Discarded plastic, paper, textiles, and glass can be recycled or reused. Companies now "mine" wastes for valuable rare earth metals and generate energy through thermal or biological treatment. An artisanal furniture maker can collect timber from demolished houses to make into furniture, a process known as "upcycling". At the same time, a large mining company can smelt gold and other metals out of discarded materials extracted from a waste dump or at an electronic waste disposal facility.

The idea of wastes as *commodities* is closely related to their role as resources. As commodities, items are pulled from the waste stream, refurbished, and resold. To give an example, a discarded iPhone might be dismantled, its valuable metals extracted, and the rest thrown away. In this sense it is a resource. It could, alternatively, be repaired and refurbished. In this sense it is a commodity. Used or second-hand goods, like clothes or cars, are commodities sold at home (after repair or refurbishment) or shipped across borders.

As resources and commodities, wastes are or become things that people buy and sell, whether locally or across borders, and on which jobs and livelihoods depend. Therefore, wastes are also a *livelihood*. Waste handling and repurposing provide livelihoods for millions around the world, from subsistence workers in the world's favelas, slums, and peri-urban areas, right up to the corporate boardrooms of waste management, recycling, and mining companies. Waste picking, collecting, repurposing, and reselling is dangerous, but it can be a route toward upward social mobility, as the history of trash collection shows (Zimring 2015), and can provide a higher living wage than other menial work. These livelihoods come under threat when bigger actors move in, whether local "bosses" or, which has happened lately, large multinational companies, who take over waste disposal, monopolizing the extraction of value from landfilled wastes.

Most recently, the emerging concept of a global circular economy, an opposing vision to the traditional "make, use, dispose" linear economy, is reshaping perspectives yet again. Waste becomes a critical *input* in a circular economy, one where everything (or as much as possible) gets cycled back into the production process, or reprocessed and recycled. Building a circular or Zero Waste economy is the goal of many campuses, cities, countries, and global actors; we discuss this perspective further in chapter 2.

Box 1.1: Perspectives on wastes

Wastes as externality: Wastes are unwanted but unavoidable byproducts of production or consumption that must be collected and discarded.

Wastes as risk or hazard: Wastes pose environmental and/or health risks, through characteristics (if it is toxic, flammable, corrosive, or reactive according to the US Environmental Protection Agency), management, or accumulation.

Wastes as commodity: Discarded objects bought, sold, or used; for example, glass, old clothes, or used cars.

Wastes as resource: Wastes are repurposed, mined, recycled, and sold/used; for example, wood upcycled to furniture or gold extracted from electronic wastes.

Wastes as livelihood: Wastes, through collecting, recycling, or dismantling, provide income to people, often informal workers, living on or near the waste, municipal employees, and some of the world's largest utility providers.

Wastes as input: Wastes that result from linear processes of production and consumption should be cycled back into production, with the goal of creating a closed-loop or circular economy.

Outline of the Book

Chapter 2 ("Understanding Wastes") lays the groundwork for later case studies. It explores how wastes are defined and described. It identifies three types of value associated with waste that connect with the six perspectives: value from waste collection and disposal, value from resource extraction, and the value of waste prevention. This chapter weaves the discussion of values in with descriptions of wastes as objects and as streams, and methods of waste collection, disposal, recycling, and reprocessing as they have evolved over time.

Chapter 3 ("Waste Work") turns to wastes as a livelihood, for multinational corporations to waste pickers. Using examples drawn from formal and informal waste economies, it demonstrates how local waste work is embedded in a wider global waste economy. It examines the emergence of transnational waste picker alliances and their role on the global stage, and the formalization of waste work in developing countries, along with possible "deformalization" in developed countries.

Chapters 4, 5, and 6 address three case studies: discarded electronics, food waste, and plastics. All three are highly visible global categories of waste that have made headlines in the past 10 years. Because they are often thought of (not always accurately) as consumer wastes first and foremost, policy prescriptions have often focused on changing the behavior of end users. These chapters take a different perspective, examining instead these wastes' identities as globalized resources or commodities, often shipped or traded across national borders.

These chapters highlight waste as a global resource frontier. They describe the risks associated with each, especially given the quantities of these wastes that are produced and shipped, and at resulting governance challenges and innovations. They highlight the importance of looking at the full commodity supply chain, from design, to sale, use, discard, and afterlife to prevent wastes, but also to fully exploit the value of discarded items while minimizing costs. Finally, they highlight the role of activism in galvanizing action and governance innovation across these complex cases.

Discarded electronics or e-waste (chapter 4) are prime examples of resource extraction. They contain many valuable materials as well as ones that are dangerous. The trade in e-waste is well known but often oversimplified as a case of the wealthy global North dumping its trash on the poorer global South. Looking at the extractive value of e-wastes, and therefore the demand for them in importing markets, helps explain why the e-waste trade has not been stopped and why e-waste trade among Southern countries has outstripped North–South trade. It shows how assumptions about what happens to the wastes when they reach Ghana or China may be wrong, but also highlights the critical importance of product design and ease of repair (and the role of the "right to repair" movement).

Food waste (chapter 5) is rarely looked at through a global political economy lens. It is important as a "new" sort of waste that quickly gained political salience in the late 2000s, and as a cross-scale problem, from individuals and households to agri-businesses. Governance initiatives have emerged at all these levels, though mostly in developed countries. It is an arena characterized by new and exciting forms of activism. But rarely are direct connections made between the shipment of agricultural surpluses as food aid from the US and EU to countries in Africa, Asia, and Latin America, and food waste discourses. Similarly, it is possible to draw direct connections between food exports and food security that challenge conventional assumptions about food loss in the production process.

Plastic scrap, waste and debris re-defined public debates over waste and its global impacts (chapter 6). Plastics clogging the oceans and killing wildlife generated a powerful activist movement in 2017 and 2018. Studies revealed how much plastic we have produced and where it has gone. This chapter turns a global resource lens onto this crisis, focusing on China's Operation National Sword that shattered an entire global scrap market, and reshaped the rules about what households could put into their recycling bins. Plastic scrap sits right at a border between "waste" and "scrap." It is hard to recycle into a reusable form, but can be recycled under certain conditions, and, given the amount of plastic waste we produce, should be reused in some form. Is this finally the point where the international community will draw a dividing line between what has value and what does not?

The conclusion draws together insights from earlier chapters. It discusses the three themes of the book: waste as global resource frontier, magnified risks, and governance challenges. It takes on a question that many in the waste and resources world are grappling with: is it possible to create a world without waste? In doing this it assesses the debate around building a global circular economy.

The Wider Significance of Wastes

In September 2018, a special report on waste in *The Economist* ("A Load of Rubbish," in the September 29 issue) marked wastes' emergence as a significant global issue. It touched on many of the topics further explored in this book: how much waste is produced globally, waste picking in the global South, the changing role of China, ocean plastics, and the dilemmas of designing and implementing circular economies.

Studying the production, management, disposal, and reuse of wastes tells us not only about wastes themselves, but also about the political, social, and economic contexts in which they are embedded. It reveals environmental injustice, and often, instances of corruption and crime. The quality of municipal waste management and collection tells how well a city is managed. Studying wastes' journeys from local nuisance to globally networked commodities illustrates how the local and the global are connected in the modern economy. They show the evolution of corporate power in a global marketplace, the emergence of transnational labor activism, and governance at multiple levels in the global system, from cities and communities on up to the United Nations.

In this book, I draw on insights from broader academic fields (see Selected Readings). The field of Discard Studies (www.discardstudies.com) looks at "wider systems, structures and cultures of waste and wasting," utilizing multiple perspectives across social sciences, humanities, arts, and science studies (Liboiron 2018). Work in the field of global environmental politics and governance illuminates how and why wastes move across borders, and existing and new options for effective global governance. These perspectives are all brought to bear to illuminate how the waste economy has been reshaped in recent decades. The chapters that follow delineate the features of this new global waste economy: what it looks like across countries and communities, who benefits, who bears the cost, who governs, and how.

CHAPTER TWO

Understanding Wastes

In 2017, in the wake of a string of deadly hurricanes across North America and typhoons in Asia, fires in California and Portugal, and the ongoing destruction from the war in Syria, "disaster waste" on a large scale started to grab headlines. It is made up of rubble, but also materials of value (copper, steel, surviving household items), and hazardous wastes (leaked oil, raw sewage, decaying food). Disaster waste also includes human remains, which make its disposal even more difficult and sensitive. As climate change (and conflict) escalate, disaster waste across the globe will pose considerable challenges in recovery processes that can last years, yet governments and relief agencies are only now starting to pay attention to it. Disaster waste embodies wastes in all their complexity. It shows the shifting terrain of wastes as a policy arena. No two disasters are the same, but there will always be value to recover and damage to mend in afflicted areas.

This chapter focuses on understanding wastes: how they are defined, how they are classified, and why they have value. It also introduces wastes and waste management in practical terms: what are the main types of wastes? How much is produced? And what are the ways they can be disposed of? This chapter identifies how value is derived from wastes, and the controversies around these processes. It notes the contingency of wastes' values in rapidly changing global markets.

What are Wastes?

> What then is waste? Is there such a thing or are there only
> different wastes? Is the difference in degree between a blob
> of mustard on a dinner plate and a worn out nuclear reac-
> tor so great (in terms of size or hazard) that it becomes a
> difference in kind? Or are there features common to all
> wastes that not only justify one designation but also suggest
> a common solution to the problems they face? (Gourlay
> 1992, p. 20)

Defining waste with precision is extremely hard, perhaps
impossible. In many jurisdictions, its definition for regulatory purposes is subject to intense legal contestation.
The US in particular has a highly adversarial regulatory
system that has been brought to bear on waste definition
and regulation (Kagan 2001). There is very little agreement
across or within countries as to common definitions of
wastes. Nor is there a widely accepted distinction between
waste and scrap or resource. These inconsistencies complicate and obstruct efforts to enhance transparency about
wastes and their movements and to improve global governance. Box 2.1 offers a series of definitions from influential
sources, which, on careful reading, differ significantly.

Each definition incorporates different understandings
of whether what we throw away still has value that can
be reclaimed or is at the end of its useful life. The *Oxford
English Dictionary* does not leave room for post-discard
use, for example, and neither does the World Health
Organization (WHO). At the other end of the spectrum,
"matter out of place," the definition of waste that has been
so influential in the field of Discard Studies (see below)
emphasizes the very contextual nature of dirt (discards,
waste). Ken Gourlay, who wrote on the global politics of
waste in its early days, captures in a clear and elegant way

Box 2.1: Definitions of waste

- "Refuse matter; unserviceable material remaining over from any process of manufacture; the useless by-products of any industrial process; material or manufactured articles so damaged as to be useless or unsaleable" (*Oxford English Dictionary Online*)
- "'Wastes' are substances or objects which are disposed of or are intended to be disposed of or are required to be disposed of by the provisions of national law" (Text of the 1989 United Nations Basel Convention on the Control of Transboundary Movements of Hazardous Wastes and Their Disposal)
- "Something, which the owner no longer wants at a given time and space and which has no current or perceived market value" (World Health Organization)
- "Unwanted or discarded materials 'rejected as useless, unneeded or surplus to requirements'" (Wilson et al. 2015, p. 22)
- "What we do not want or what we fail to use" (Gourlay 1992, p. 21)
- "Matter out of place" (adapted from Mary Douglas's classic definition of dirt in Douglas 1966)

Adapted from Gourlay (1992)

the contextual dimensions of waste. His definition assigns responsibility ("we," whoever that may be), while opening the door to a useful afterlife for "what we do not want or fail to use."

This point – that things that have been discarded can be both wastes and resources – is encapsulated in the many different words for "waste" in figure 2.1. Some of these words represent the end of life perspective, others refurbishment, reuse, and resale.

Given the rich historical lexicon of words for waste, detritus, litter, etc., I have surely left many colorful terms out of this word cloud, which came out of several years' research and reading. I weighted the terms myself. "Scrap" is the second standout term because distinguishing between it and waste – with and without perceived value – is at the heart of global disputes over the international trade in

Figure 2.1: Words for waste

Source: Author, created using WordItOut.com

wastes and scrap. Scrap is the generic term for used goods and materials that can be resold and remade into something with value. In the US, for example, there is a clear demarcation of what constitutes "scrap" for the purposes of the market, but this kind of distinction does not hold worldwide, creating legal, political, and cultural gaps that are hard to bridge (see chapter 6).

Some advocates of the global circular economy suggest avoiding the term waste and its synonyms altogether (see the book's conclusion). Using terms like used, second-hand, or secondary helps shift the paradigm away from seeing discarded goods as "end-of-life," certainly for consumer goods. Chapter 4, on electronic waste, demonstrates the power of this distinction, and how it is being used to assert the right to repair broken electronics and appliances.

At the same time, not all waste can be reclaimed, and some remains dangerous and unusable, such as medical wastes from radiation or chemotherapy. Some materials can only be repaired or recycled a few times before they become useless: plastic, for instance, downcycles quickly. Resources exist only when there are people ready to use or buy them. For instance, it is hard to find markets for nuclear waste, raw sewage, or toxic ash, at least for now. It remains important to choose terminology that does not gloss over the very real shadows of waste production and recycling.

Tracking Wastes' Journeys: Streams and Objects

This book's primary empirical cases – electronic, food and plastic wastes or scrap – are complex examples of how the new global waste economy works. They can be viewed as *objects*, or things, for example, cellphones, leafy greens, and plastic straws. Their journeys can be followed from "birth" to afterlife. And they are waste *streams*, containing items of value (gold and metals, recyclable plastic, compost), and risk (mercury and dioxin, contaminated foodstuffs). This section introduces these two ways of thinking about wastes (scrap, etc.) and their lifecycles. While we often think of wastes as *objects*, recognizable things or commodities, they are also defined in terms of *streams*. These two ways of conceptualizing wastes help govern how wastes are produced, handled, disposed of, or sent back into use.

Wastes as Objects

One way to understand wastes-as-resources is to see them as *objects*. Electronics, batteries (from AA household batteries to those that power hybrid and electric cars), tires, plastic water bottles, straws and coffee pods, second-hand cars, and used mattresses and clothes all cut across standard categories of waste. They contain organic, metal, plastic, recyclable, non-recyclable, hazardous elements, or some combination. When we use a product, it is easy to think only about the use that *we* get out of it and ignore its life before we buy it and after we throw it away. This is even true of the most ubiquitous commodities, such as the clothes we wear.

These discarded objects do not always vanish. Often, they are recognizable when they show up in landfills, on pavements, and in scrapyards. The rusted bodies of discarded cars used to litter the US landscape. They show up as props in movies: Stephano Bloch has written about the after-life of mattresses dumped on the streets of Los Angeles (Bloch 2013). Several years ago, I was doing a search of images of end-of-life ships and, by strange coincidence, happened across the ship that had taken my dad to England in 1961 and my newly-wed parents back to Australia in 1965. It was virtually intact, but trashed, in a marine scrap yard in China. I recognized the name, *The Oriana*, sent the photo to my dad for confirmation, and he was able to point out their cabin.

Tracing the journeys of "things" like T-shirts from points of discard demonstrates how they acquire, lose, and re-acquire value over time. Such analyses also increase the visibility – and therefore, the broader resonance – of wastes as complex social, economic, and governance problems. Further, tracing their journeys from the point of their *production* through point of final disposal connects their environmental, social, political, and economic aspects across entire product lifecycles.

Many used or second-hand goods are traded across borders, notably used electronics, which may be repaired and reused, or they may be dumped. We look at them in depth in chapter 4. Thriving export-import businesses exist for trading used cars (and other modes of transportation, such as bicycles). Brooks (2012) explores the networks of actors and networks involved in shipping used cars from Japan to Mozambique via South Africa by Pakistani traders. Millions of used cars are exported from the US to Mexico each year, made easier under the North American Free Trade Agreement (NAFTA). Russia and Burma are also lead importers of used cars, according to one 2016 study (Coffin et al. 2016).

Shipments of used clothing to developing countries are another high-profile example (Brooks 2013). Even in the mid-1990s, donations were overwhelming demand for used clothing within the US, forcing charities and clothing collectors to seek other options. A 2017 report from the Ellen MacArthur Foundation (*A New Textiles Economy: Redesigning Fashion's Future*) found that of the clothing collected for reuse in developed countries, up to half is exported to poorer countries. Global clothing production doubled in terms of units between 2000 and 2015, to 200 billion units. Clothing has also declined in quality with the rise of "fast fashion," clothes designed to be worn only a few times, then thrown away.

African countries, overwhelmed by "donations" that threaten their domestic textiles and clothing industries, have started to refuse shipments. Members of the East African Community (including Rwanda, Kenya, Uganda, Tanzania, and Burundi) plan to ban second-hand clothing imports by 2019. Other countries, such as Ghana, focus on importing (for sale) only the highest quality clothing (Postrel and Minter 2018). Whether this feeds back into better clothing recycling, manufacturing, and design for developed countries and the countries where the clothes are produced, or greater pressures to maintain this "trade," more accurately described as dumping, remains to be seen.

Wastes as Streams

A more complicated way of looking at waste is to see it as a stream. A waste stream is the flow of waste from its point of production (or discard) through to final disposal. Waste streams are defined according to point of origin and content, as described in box 2.2.

Municipal solid waste is the most visible waste stream in developed countries, although it is dwarfed in volume

---

Box 2.2: The major waste streams

Sewage: Water-carried wastes, including greywater (water from baths, showers, dishwashers, and other appliances), human bodily wastes, soap, and detergent.

Municipal solid waste (MSW): Waste collected and treated by or for municipalities. It includes post-consumer waste from households, commercial entities, office buildings, hospitals, small businesses, and government agencies, including organic wastes, plastic, paper, glass, and aluminum. Plastic and paper wastes include bags, bottles, industrial plastics, newspapers, scrap paper, paper plates and napkins, cardboard boxes, and other packaging materials. MSW also includes yard and garden waste, street sweepings, and the contents of litter containers. It may also contain small quantities of hazardous wastes. MSW makes up about 24% of total solid wastes in OECD countries.

Industrial Waste: Wastes from manufacturing processes and production, including chemicals, iron and steel, pulp and paper, plastics, glass and concrete, food, and solid wastes from energy production. This is the largest waste stream by volume.

Agriculture and Forestry Waste: Agricultural waste includes manure and other wastes from farms, poultry houses and slaughterhouses; harvest waste; fertilizer run-off from fields; pesticides that enter water, air or soils; and salt and silt drained from fields. Forest waste also includes significant quantities of biomass (wood chippings, tree trimmings, and other waste from forest/plantation management).

Construction and Demolition Waste (C&D): This consists of many materials, including concrete, bricks, gypsum, wood, glass, metals, plastic, solvents, asbestos, and excavated soil, many of which can be recycled. It arises from activities such as the construction and demolition of buildings and civil infrastructure, and road planning and maintenance. C&D wastes are the largest waste stream in terms of weight, also voluminous, making up 34% of urban waste in OECD countries. High construction countries such as China and Abu Dhabi are rapidly catching up.

Mining and Quarrying Waste: This consists of wastes from the extraction and processing of mineral resources, including topsoil, rock, and tailings. Most mining waste is inert and poses a threat primarily through its sheer volume. But other mining residues can be highly dangerous, given the chemicals used in processing ores. Mercury, for example, used to process gold ore, continues to be present around active and abandoned mines.

Hazardous Waste: Waste is hazardous if the risks it poses to human health and environments over the short and long term need it to be handled and disposed of using specialized methods. It includes byproducts of manufacturing processes, medical procedures, and discarded commercial products, like cleaning fluids or pesticides. Household hazardous waste includes paints and solvents, asbestos, discarded medications, and e-waste, including fluorescent light bulbs and cathode ray tubes (from old TVs), and batteries. It includes stockpiles of wastes from decommissioned military bases or dismantled chemical or biological weapons.

Nuclear (Radioactive) Waste: This consists of spent fuel rods from nuclear power reactors, waste from the production or reprocessing of nuclear weapons, and certain kinds of medical waste or from research experiments. Due to the extreme risks it poses, its lengthy half-life, and security risks, nuclear waste is managed by government agencies, and subject to strict controls.

Sources: Wilson et al. 2015, OECD data on MSW (at https://data.oecd.org/waste/municipal-waste.htm), and the European Commission (at http://ec.europa.eu/environment/waste/index.htm)

and weight by industrial and construction and demolition wastes. Available data suggest that globally, the amount of industrial wastes produced annually is more than 18 times greater than MSW (Kaza et al. 2018, p. 37; see also Liboiron 2016; MacBride 2012). They often contain higher volumes and more readily extractable resources – such as copper piping, or steel support beams – but MSW is more easily accessible to waste pickers and others who seek to exploit its value.

Globally, the largest waste category is green waste, including food, at 44 percent (Kaza et al. 2018, p. 29). "Dry recyclables" – plastic, paper products, metal, and glass – add up to another 38 percent (Kaza et al. 2018, p. 29), with green, or organic, wastes making up a much higher proportion of the waste stream in low-income than in high-income countries. These ratios are expected to change with growing rates of urbanization in the global South; in fact, between the 2012 and the 2018 versions of "What a Waste," the proportion of organic waste in the global waste stream was estimated to have dropped from 64 to 56 percent (Kaza et al. 2018, p. 2).[1]

These streams do not stay in neatly separated channels. Hazardous waste – also known as toxic, special, or scheduled – finds its way into the MSW stream, as medical waste, for example, or batteries. There are other cross-cutting waste types. These include "mega-wastes" (end-of-life ships or oil rigs), and "wastes from waste" such as incinerator ash, often highly toxic, and disaster waste.

The concept of streams is a useful metaphor for understanding resources and risks. Resources and reusable commodities are removed or extracted from the flow and returned to economic circulation: metals, compostable food, or hard plastics (to draw on our cases). What's left in the stream to flow on to final disposal, are the "unusable" (less usable) components of the stream (depending on cultural context) and the most hazardous, such as lead or mercury, diseased meat, or plastic film. Dealing with the massive volumes of MSW produced globally means thinking not only about plucking items out of the stream, but diverting it altogether, or reducing its flow. For example, "Zero Waste" often means, in practice, zero waste diverted to final disposal in landfill (see box 2.5). Waste can be diverted to reuse or recycling, or, more controversially, to energy production. Sweden has done this successfully with its waste prevention and waste-to-energy programs (Yee 2018), to the extent that it now needs to import waste from its neighbors to keep the fires burning. It is, however, finding disposal of its toxic ash ("waste from waste") problematic. Waste prevention measures – through better product design, for instance – reduce waste streams' flow in the first place.

"Where there's Muck, there's Brass"[2]: Perspectives on the Value of Wastes

As the preceding sections show, wastes are not a worthless drain on our economy or society. They are far more complex and valuable than that. Value exists in almost every type of discarded material, even the waste we might most want to disappear. Treated sewage becomes fertilizer. Ambergris, used historically in perfume manufacture, is excreted by sperm whales. Guano is produced by sea-birds in such quantities on islands in the South Atlantic and Eastern Pacific oceans that it has been mined for fertilizer production.

The business of creating value from wastes extends from the very biggest global corporations to your neighbor's hobby of fixing old cars. These activities occur in every part of the world. They engender large transboundary flows of (trade in) scrap, products, and technology. They are undertaken by some of the world's largest multinational corporations in the waste, energy, and mining sectors. In response, transnational activist movements have started to advocate at a global level the interests of informal workers – and local communities – against threats to their livelihoods, health and safety, in part because of the unevenness and contingencies associated with transforming wastes into objects of value. As chapter 3 shows, millions of people in rich and poor countries depend on waste, what we do not want or cannot use, for their livelihoods, and much money is made in the waste economy.

Still, in many cases the damage imposed by wastes themselves (e.g. toxic wastes) or by badly managed wastes, far outweigh any value that might be derived. Also, this analysis by no means implies a license to freely produce wastes. Quantities of waste are overwhelming ecosystems (as in the case of plastics in the ocean) and we do not have fully functioning, socially equitable systems to extract all possible value from the wastes we produce while minimizing the harm to workers.

Three sources of value (money or profit, namely, the "brass") in wastes are particularly pertinent to this narrative: value from waste collection and disposal, extractive value, and value from waste prevention and diversion. Each maps onto one or more of the perspectives discussed in chapter 1: wastes as risk or externality, resource, commodity, livelihood, and input.

However, wastes' translation into resources may be partial, temporary or fragile; in other words, contingent. Wastes' value depends on exogenous factors, and the perceptions of buyers and sellers. As secondary resources, their markets are dependent on markets for primary resources. Exploitation of wastes can leave behind a literal wasteland, damaged human health and damaged environments. That the business of waste disposal and reprocessing is often relocated to poorer parts of the world magnifies these impacts.

Waste Collection, Removal, and Disposal
First, waste has value for those who collect, transport, and dispose of the wastes. Someone is paid to take it away and deal with it through proper disposal and management. In the simplest sense, this is money in the pocket of those who treat waste as an end-of-life substance and provide waste management as a service, and a benefit to those who produce the waste in the first place. Government agencies may also benefit from fees and taxes, such as landfill taxes,

> ## Box 2.3: Waste collection and removal
>
> **a. Not collected:** Waste and litter left on street, in neighborhoods, or left behind after manufacturer or other entity moves out.
>
> **b. On-site management:** Some wastes are stored, recycled/reprocessed, or disposed of on-site, including agricultural waste, mining, and quarrying wastes, spent nuclear fuel rods, decommissioned military bases.
>
> **c. Collection and transported by municipal agencies or private haulers:** Curbside (consumers put the waste out in trashcans, recycling containers) or from dumpsters (businesses, apartments, institutions). Waste collection companies are often private sector firms under contract with municipalities.
>
> **d. Collected directly from private producers:** Waste disposal companies collect industrial wastes, construction and demolition wastes, and others from the factories and other installations that produce them.
>
> **e. Informal collection:** By local waste workers, small haulage companies.

which generate significant revenues for cash-strapped authorities.

The basic principles of end-of-life waste management have stayed the same since their early days in the Industrial Revolution, though practices and technology have changed considerably. Boxes 2.3 and 2.4 outline the major features of waste collection and removal, both controlled and uncontrolled. While controlled waste management – the collection, transport, and disposal of wastes in a safe and environmentally sound manner, with oversight and accountability – is the end goal of waste management systems, this ideal is not so easily achieved.

Although waste collection services date back to Ancient Greece and the Roman Empire, the story of formal waste collection, transport and disposal in the industrialized world begins with cities and towns in Europe and the US after the Industrial Revolution. Increasing amounts of dirt and debris from industrial production, and health threats in urban areas, such as cholera, led authorities to turn attention to waste management.

> ## Box 2.4: Controlled waste disposal and storage
>
> **Landfilling** is the burial of wastes in controlled and contained sites. Wastes may be treated or untreated, unmixed, or mixed (mixed municipal, industrial and/or hazardous wastes). Sanitary landfills, lined, layered with soil or organic waste, and carefully monitored are the safest type.
>
> **Incineration or thermal treatment** produces ash, gas, and heat. It reduces wastes to remnants that can be disposed of more easily or used as fertilizer. However, smoke can be highly toxic, especially with uncontrolled/open incineration at lower temperatures. If the waste is mixed, the ash can also contain highly concentrated toxins.
>
> **Composting** is the aerobic decomposition of organic wastes, including food, animal waste, and agricultural wastes. Subsequent compost may be a valuable addition to soil. It is often very cheap and can be used at household or municipal levels.
>
> **Anaerobic digestion** is used to treat organic wastes, which decompose in an oxygen-starved environment. It produces gases that can then be used as energy. It minimizes odor and other effects of traditional composting methods, deals with wet, heavy wastes, and can be deployed at a large scale.
>
> **Long-term storage** (including for nuclear and some persistent organic pollutants) in specially constructed facilities, isolated from human settlements, vulnerable ecosystems (and, ideally, low seismic risk).
>
> **Remediation** involves cleaning up old waste dumps, decontaminating soil, water, and local ecosystems.

By the late nineteenth century, municipal authorities had taken over from informal collectors and collection organizations to establish systematized waste collection and haulage systems that would protect public health. In fact, waste management was one of the early services municipal authorities took on. Some of the companies that are global giants today had their start during these times, gaining a competitive edge through innovations, such as designated garbage trucks and bins. In the 1970s many developed country governments instituted formal recycling systems. In subsequent years, regulatory focus shifted to waste prevention, introducing the waste management hierarchy. There are various versions of this hierarchy but the comprehensive ones list, in order of preference: waste prevention, minimization, reuse, recycling, and finally, disposal.

Waste management as end-of-life disposal is designed to be an invisible service. For most services (banking, haircuts, restaurant meals), we pay money to see a tangible result. In the case of waste, we pay to never see something again. This can create perverse incentives for the waste collector, who could simply dump hazardous wastes in a landfill after being paid for specialized services, a practice well documented in studies of environmental crime (Szasz 1986). It also underlies the emergence of the global hazardous waste trade in the 1970s and 1980s.

Landfill is the most common means of end-of-life waste disposal. Fifty-five percent of waste in the US is buried in covered landfill. In the EU in 1995, the proportion was about the same, but by 2018, barely a quarter met this fate, according to 2019 data.[3] Globally, this total was around 40 percent in 2016, 10 percent via recycling and composting (a large rise in recent years), 11 percent through incineration and the remainder through open dumping (Kaza et al. 2018, p. 34).

The other long-standing technology for getting rid of waste is incineration, one of the major contributors to human health damage from waste disposal, especially for poor or ethnic communities and communities of color. In developed countries, waste incineration with no energy generation is being phased out, with few to no facilities under construction. Instead, technological innovation in incineration and biological treatments have shifted this sector into energy production, another source of value extraction (see below). China's big push toward waste-to-energy incineration under its 12th Five Year Plan drove an unprecedented increase in its overall share of all methods of waste disposal in upper income countries, from 0.1 percent in 2012 to 10 percent in 2018 (Kaza et al. 2018, p. 2).

Millions make their livelihoods from waste collection and disposal. But the story does not end there. Those who collect waste are often also those who recycle, reprocess, and resell: the extractive value of wastes.

Extracting Value from Wastes: Recycling, Reprocessing, Recovery, and more

Second, wastes have extractive value as resources and commodities. Reprocessing, recycling and repurposing wastes and discarded materials to create new material inputs and energy all add value (back) to what others have thrown away. Metals are extracted from e-waste through reprocessing (dismantled and crushed, metals separated, melted down and reused). Timber can be repurposed and reused in building or furniture. Electronics and clothes can be refurbished, while aluminum cans, glass bottles, plastic, paper, and cardboard can be recycled.

The simplest form of the waste disposal hierarchy, an educational tool for handling waste, is the "Three R's" (Reduce, Reuse, Recycle, in order of preference). In today's complex waste economy, there are at least 13 R's, which include such additional terms as Repair, Recover, and Regenerate. The double-loop Möbius strip became the official recycling logo in 1970. Transfer stations (where people bring their recyclables) were supplemented by curbside collection programs in the 1970s.[4] The first mandatory recycling program in the US was established in Woodbury, NJ, in 1980. Recycling is not always straightforward. In the best case, closed-loop recycling, products can be recycled indefinitely back into the same product or something of equal quality. But for many products, such as plastics, open-loop recycling dominates, and they are often downcycled, into something of lower quality, and may rapidly downgrade with successive cycles.

This is a very Northern experience, that recycling is something that must be managed or mandated rather than something that happens habitually, whether as a business or for peoples' own use. Jo Beall has pointed out that while archaeologists studying the twentieth century would have rich pickings in landfills in the global North, there would be very little left to pick over in disposal sites in the global South, in those areas where recycling norms remain strong (Beall 1997, p. 77).

Recycling and other value extraction activities are now carried out on a global scale, as wastes and scrap are shipped across borders for dismantling, recycling, and reuse. "Urban mining," the process of extracting or reclaiming valuable resources from existing reservoirs of waste, is carried out by small local companies and multinational corporations alike. We discuss urban mining in depth in the context of electronic wastes in chapter 4.

Another way in which value is extracted from waste is through energy recovery. Seven hundred years ago Marco Polo reported observing Chinese peasants capturing methane from covered sewage tanks to power their huts. Waste-to-energy (WTE) technologies are now far more advanced, but they are also extremely contentious, especially thermal technologies, namely, incineration.[5]

The global WTE market was worth $26 billion in 2014, projected to be $44 billion in 2024, according to data collection service Statista. The market includes landfill gas capture for energy generation as well as thermal treatments. In recent years the leading waste companies have diversified their portfolios to include large-scale or advanced WTE, and cities like Los Angeles are converting landfills to energy production for surrounding communities, such as the Puente Hills landfill, which, when it closed in 2014, was the largest landfill in the US. Several of Japan's municipal governments (Tokyo, Yokohama, Kobe, and Osaka) are among the biggest global WTE players. China has become the new global center for WTE incineration, as its waste production grows, and its landfills run out of space. Under its 12th Five Year Plan (2011–2015), 244 additional incinerators were constructed, taking in 35 percent of its waste and burning around 224,000 tons per day. By the end of the 13th Five Year Plan (2016–2020), WTE capacity is intended to double, to 467,000 tons per day (Li et al. 2015). Sweden, the poster child for WTE incineration, incinerates nearly 50 percent of its MSW to generate energy.

Supporters point out WTE's utility in countering climate change in addition to reducing global stockpiles of wastes, especially plastics (Castaldi 2014). Traditional combustion technologies are now supplemented with more advanced technologies. Gasification and pyrolysis are forms of thermal waste treatment that burn at higher temperatures, pollute less, and produce a liquid "syngas" that can be used to power vehicles. "Chemical" recycling (also known as thermal depolymerization) breaks plastics down to their monomers, their most basic components, allowing for plastics to be recycled without downgrading.

Opponents counter that even the most advanced technologies have human health and environmental costs. Incineration, with or without energy generation, has long been known for its lasting negative impacts on surrounding communities, especially through the emission of dioxins and other highly toxic chemicals (Baptista 2018). Halting the siting of incineration facilities in minority communities was a key impetus for the US environmental justice movement (United Church of Christ Commission for Racial Justice, 1987). Organizations such as the US

Council of Mayors do not recognize WTE incineration as a source of renewable energy, or as meeting Zero Waste goals. Part of the reason for this perspective is recognizing the health and environmental burdens that incineration have that counter any climate benefits; another is that it implicitly encourages continued waste generation. Anti-incineration activism is a global phenomenon. In the UK, Ireland, China, Russia, and many other countries, communities have mobilized against WTE incineration, with lawsuits and protests.[6] GAIA – a global network that mobilizes against incineration and for equitable Zero Waste policies – has long opposed incineration of waste for energy.

Waste Prevention and Diversion
A third source of value in wastes comes from benefits derived from preventing or diverting wastes, connecting to the "wastes as inputs" perspective. Waste prevention is one of the central pillars of circular economy platforms (see box 2.5). It improves efficiency, saves resources, and reduces pollution burdens. In practice, it has two aspects. Waste *prevention* (or avoidance) means producing only minimal amounts of wastes that cannot be cycled back into production. Waste *diversion* means that wastes are diverted from final disposal in landfill, open dump, or waterways toward productive uses: recycling, reuse, energy generation, compost, and so on. Waste diversion is the underlying principle of many local Zero Waste initiatives put in place since the early 2000s.

The distinction between waste diversion and prevention is important because the two approaches mobilize different ideas and constituencies. The waste industry, for example, is positioning itself as the lead provider of waste diversion services (recycling, energy, composting), pitting itself against waste prevention advocates who want to shrink waste streams in the first place.

Box 2.5 defines and identifies advocates and applications of two terms that dominate contemporary debates over wastes and resources: "circular economy" and "zero waste" capture visions of local and global waste economies that integrate the preventive and extractive values of wastes. Later chapters evaluate the strengths and weaknesses of these visions, especially when they are scaled up to the global level. While food waste has provided fertile ground for enacting circular economy policies, plastic wastes are proving extremely challenging. For example, discarded food can be collected and composted or redistributed locally. Plastics, on the other hand, are much harder to recycle locally, are often shipped long distances, and are much harder to recycle into something marketable.

Reducing the amount of waste produced or in landfills is sorely needed for many reasons. The global cost of dealing with the world's rapidly accumulating municipal solid waste will rise from $205 billion a year in 2010 to $375 billion by 2025 (Wilson et al. 2015). These totals do not include the cost of dealing safely with hazardous or nuclear wastes. Around the world, collection, recycling, and disposal infrastructures, already inadequate in most of the world, are strained beyond capacity. Space to store this waste is running low, even with the inventiveness – and carelessness – humans have demonstrated to find places to put waste.

Actual monetary flows in this case are more complicated and harder to trace, over and above the profits derived from recycling and resource extraction. They exist primarily as money saved through higher levels of efficiency. The main value in waste prevention or diversion lies in avoided human health impacts, environmental contamination, and greenhouse gas emissions. It also lies in avoiding

Box 2.5: Zero Waste and the Circular Economy

The terms Zero Waste and Circular Economy describe policy initiatives, grassroots campaigns, entrepreneurial activities and industrial and product design that move beyond the linear "make, use, dispose" economy.[7] For some, these concepts represent a new global aspiration or vision: a planet where there is no such thing as waste. The distinctions between them are not hard and fast, but they are often used in different contexts.

Zero Waste (ZW) as a policy goal is adopted primarily at local levels, by counties, college campuses, and cities such as San Francisco (full zero waste by 2020), London (65 percent zero waste by 2030) and Kamikatsu, Japan (full zero waste by 2020). Kamikatsu is already known for requiring citizens to sort their trash into 34 categories. Definitions explicitly stress the goal of diverting waste from landfill and incineration, but often extend to embrace product and packaging re-design and use. Activist groups such as Zero Waste Europe have espoused zero waste for grassroots mobilization. The term is used to express practical and achievable goals. This definition from the Santa Ynez Band of Chumash Mission Indians, from a US EPA webpage on how communities have defined Zero Waste, clearly and concisely captures the concept: "to reduce the waste that goes to landfills and incinerators to as little as possible (zero is the goal) and to redesign products, packaging and other items so that they can be reused or otherwise avoid the landfill" (US EPA n.d.).

The concept of the *Circular Economy* (CE) captures a larger vision of a global sustainability transition. The UK group WRAP's definition is widely cited: "[A]n alternative to a traditional linear economy in which we keep resources in use for as long as possible, extract the maximum value from them whilst in use, then recover and regenerate products and materials at the end of each service life" (WRAP n.d.). It has been adopted by the EU and China as an underlying development paradigm for the twenty-first century (Kiser 2016; Mathews and Tan 2016; Stahel 2016). The Ellen MacArthur Foundation (www.ellenmacarthurfoundation.org) has taken as its mission to accelerate the global "transition to a circular economy." Their consulting, research, and advocacy work has defined the field.

Implementing CE platforms means redesigning products and supply chains and shifting to renewable energy and sustainable agriculture strategies across economic systems. Changing how we produce and dispose of wastes so that, in the ideal world, there is no end-of-life disposal, is central to circular economy platforms. By extension, food, electronics and, most particularly, plastics are critical components of these plans. The EU's 2018 CE Package puts plastics front and center, with ambitious goals to dramatically reduce single-use plastic consumer goods by 2030, mandate producer take-back and recycling of discarded plastics, especially packaging, and measures to cut the amount of plastic entering the ocean.

legal liability for accidents and contamination. Economic models can evaluate these cost savings in monetary terms. There are also clear gains for entrepreneurs and designers, consultancies and researchers providing ideas, technology, and expertise. There is a thriving circular economy startup sector, with entities such as Closed Loop Ventures providing initial capital, and awards to win (such as Europe's Green Alley award for circular economy startups). Financial (as well as normative) incentives to design wastes out of the system are many, including the possibility of patented projects.

Zero waste and circular economy (CE) policies and programs provide opportunities for firms that can invest in advanced WTE or materials recovery and recycling technology. There are government mandates or programs for all or select waste streams in the EU, in US states such as California and Massachusetts, and cities around the world. Food waste is one (see chapter 5). For the industry, making these shifts, and making money while helping divert waste away from straight landfill to more productive uses, is central in their positioning as suppliers of environmental services. Zero waste goals set by cities, campuses, and others are ambitious but can be misleading or poorly designed. If zero waste simply means no diversion to landfill from the immediate point of disposal (such as into household recycling or green waste bins), programs neglect what happens to the bin contents after collection. That so much discarded plastic was sent to China post-collection and sorting in industrialized countries while being counted toward diversion goals was disillusioning (Davis 2014).

Despite its normative appeal as "technologically driven and economically profitable" (Hobson and Lynch 2016, p. 15) and its support at the highest levels of government around the world, the CE idea has been strongly criticized in theory and in practice, and indeed the term obscures some very real conflicts. One is highlighted in subsequent chapters: the role of global trade in e-wastes and plastic scrap. This trade has underpinned recycling systems in all developed countries. Simply ending waste exports from developed countries (as CE advocates suggest) would disrupt recycling economies and the use of recycled material in manufacturing markets (Gregson et al. 2015), even as we must take account of the environmental and social injustices inflicted. Another conflict is over whether WTE incineration should be considered part of a CE: most activists and many policy makers say no, while the WTE industry and China say yes. The recent trend has been to exclude incineration from official Zero Waste and Circular Economy policy platforms. We return to these debates and critiques in the concluding chapter.

The Non-Material Value of Wastes

Not all the benefits derived from waste are monetary. Art made from discarded objects and landfill detritus, food waste pop-up restaurants, and musical instruments found or fashioned from landfilled material enhance communities and create new cultural experiences. The "Recycled Orchestra" of Cateura, Paraguay, was the subject of a 2015 documentary, Landfill Harmonic. End-of-life ships sunk for coral habitat show that wastes provide ecosystem services. Plastic waste sculptures in museums, in city squares, and at global political meetings draw attention to this problem in an eye-catching way. Galleries in New York City and elsewhere have hosted interactive e-waste art exhibitions. These approaches blend utility, art, and politics.

Wastes also tell us about our everyday history. Ancient waste repositories, or middens, are an invaluable source of historical information about the lives of our ancestors. Landfills in place for 100 years or more are the subject of "garbology" (Humes 2010). Taking a core sample from such a landfill tells us what people were eating, reading, and using at different times. This work has also revealed that industrialized country trash does not decompose easily under landfill conditions.

Wastes as Contingent Resources

Despite all the ways wastes contain value, their transformation from something to be disposed of to a valuable resource or commodity may be partial, uneven, or temporary – in other words, contingent. Their value is uncertain, unpredictable, and contextual, dependent on many external factors. It is often trapped, literally (surrounded by hazardous materials), or figuratively, as markets fluctuate. Cardboard packing boxes (see chapter 1) are valuable to some, trash to others. Spent nuclear fuel rods can be reprocessed back into valuable fuel for powerplants, but still ultimately leave radioactive waste that must be stored for hundreds of thousands of years. Such contingencies have the potential to generate additional risks and governance challenges.

Markets for wastes and recycled goods are vulnerable to economic and price shocks (Jolly 2007). The profit margins for scrap are low, so even slight shifts will have big impacts on market strength. Markets for secondary materials are highly dependent on the price of the equivalent virgin materials. If prices of oil or copper fall significantly, whole markets for plastic or copper scrap could collapse, leaving piles of unwanted materials lying in scrap yards or municipal sorting areas. In the absence of someone to sell them to, holders of scrap will look for the cheapest disposal route, often the least environmentally sound. In turn these fluctuations have deep impacts on the people who depend on resource and commodity extraction from waste for their livelihoods.

Some discarded items, such as scrap metals or aluminum cans can be turned back into something of the same quality, if they can be captured by the recycling system. This is called closed-loop recycling. Others, however, can only be downcycled, recycled, or reprocessed into something of lower quality, until they degrade and become unusable (called open-loop or cascaded recycling). Certain plastics reach a final use point quickly.

Other wastes when reprocessed or recycled generate residue that cannot be reprocessed any further. The ash from WTE processing is an example of concentrated waste that contains many of the toxic components of the substances that were burnt. The disposal of these residues may further impact local environmental injustices in communities around WTE incineration facilities. Some waste streams become too costly to sort effectively: as municipal waste streams become more contaminated with medical wastes, electronics, batteries, and the like, the entire stream is more likely to be sent to landfill.

Conclusions

This chapter has introduced examples and categories of what we do not want or fail to use. In thinking about wastes as both objects and streams, it is possible to connect them with value, particularly value beyond disposal. The "wastes as objects" perspective connects to the full life cycle of products, from initial harvesting of resources, to manufacture and use, to final disposal. "Wastes as streams" focuses attention on what happens post-discard, given that the act of discarding itself is far from end-of-life.

The new global waste economy has transformed the governance landscape. To return to Gourlay's question quoted at the start of this chapter, "are there features common to all wastes that not only justify one designation but also suggest a common solution to the problems they face?", increasingly the answer would be no. There are a large category of objects and streams that at some point after they have been used can all be called "wastes" or "discards" but this is far from their whole story.

Waste regulations and governance that focus solely on "end of pipe" solutions (governing waste as an externality or risk) or at the local level are unsuited to this new global waste economy. Likewise, existing global agreements and institutions are weak and ill-equipped to deal with new types of wastes and waste flows, and with wastes as a resource rather than risk. The entire architecture of collection, disposal, and recycling is not enough to handle the global waste crisis. Further, little attention is paid by aid agencies to the need to address this crisis: only 0.3 percent of international development financing goes to solid waste management, and of that two-thirds went to ten middle-income countries (Wilson et al. 2015, p. 204).

Waste governance should be about minimizing risks to marginalized workers and communities, about creating and maintaining strong, stable, and safe national or global markets for secondary materials, and implementing shifts to more efficient, more circular economies. But this can only be done based on understanding that wastes are important resources, and that they have value. That way, waste governance measures will gain support from more diverse actors (including industry) and be more beneficial to the global economy.

The following chapters address resources, risk, and governance on this globalized frontier, and the politics of meeting these challenges. Chapter 3 addresses capital, labor, and technology in the global waste economy, the messy business of separating the brass from the muck.

CHAPTER THREE

Waste Work

Each time my fellow waste researcher Raul Pacheco-Vega visits a new city or town he looks for who is doing waste and recycling work out on the streets. Who are the people with carts full of scrap or discarded goods? Or aluminum cans and glass bottles that have a small return value? Who is sweeping the streets? Or selling second-hand books or household items on the pavements? What is their gender, age, ethnicity, or race? Are they immigrants? Migrants from rural areas? Once you start, it is hard not to see (and smell) this world. You can also look at the company names on the garbage trucks – Waste Management or Veolia (or, in parts of Chicago, Groot), and at who drives them, and who loads the garbage into the truck. Where and how are repair services offered? These are the most visible representatives of a vast industry of formal and informal work. It is harder to see or visit the waste pickers working on landfills or sorting and cleaning recyclables in their homes, or the workers operating laser sorters in materials recovery facilities (MRFs) or implementing the latest recycling technologies.

Everyone lives within a local waste economy. In recent years, whenever I moved to a new house or downsized possessions, I hired a local "man with pickup truck," a retiree from the US Environmental Protection Agency (who is now retired from the junk business). He would take and sort all our unwanted stuff: refurbishing, recycling, selling, donating, and disposing of it. These are things I had neither the time nor the relevant knowledge to do, nor would the city pick them up. We would often chat about the rich-country folly of renting a storage unit for years then throwing everything out. I took books, batteries, old clothes, large cardboard, and other stuff that could not be put out on the curb to the city recycling center. Ads for unwieldy furniture for free pickup went on Craigslist.

Our local waste economies are also embedded in the *global* waste and recycling economy. Discarded goods and trash traverse the planet and multinational corporations in the industry invest in waste services in countries from Europe to Africa. This global economy has only recently been made visible in all its diversity. Within the global waste economy, it is also hard to track related flows of money, between whom and how much, although it is easier to see who bears most of the risk.

This chapter is, therefore, about the complicated business of extracting value from waste and discarded goods, and the people and companies whose livelihoods depend on waste. From corporations headquartered in Houston or Paris to workers' collectives on the streets of Cairo or Belo Horizonte, Brazil, millions do this work. The chapter focuses specifically on the informal sector of this vast workforce, and how it has evolved in an era of globalization. Conflict occurs when waste pickers (informal laborers) are displaced from the site of their livelihood as their governments allow formal sector corporations in to manage waste disposal sites. Waste work is often highly hazardous, and those risks have magnified as wastes increase in volume, become more toxic, or travel further. But globalization has also provided unexpected spaces for activism and technological innovation, and the spread of new organizational models for waste collection and management. The chapter uses two cases – the Zabaleen in Cairo and informal waste pickers attending global climate meetings – that demonstrate how local waste work is embedded in national and global economic and political processes. The overarching narrative takes us from the formal sector to the informal, and to the formalization and deformalization of waste work. This labor also makes resource extraction from wastes and discarded goods possible, a theme continued in chapters 4 and 6.

Waste Work and Livelihoods

What does it mean that waste is, beyond being a resource or a risk, a *livelihood*? Livelihood, as a concept, is more than just income generation. It is "gaining and retaining access to resources and opportunities, dealing with risk, negotiating social relationships and managing social networks and institutions within households, communities and cities."[1] Studies of waste work – often based on direct observation and participation – such as Robin Nagle's *Picking Up: On the Streets and Behind the Trucks with the Sanitation Workers of New York City* (2013), and Katherine Boo's *Behind the Beautiful Forevers: Life, Death, and Hope in a Mumbai Undercity* (2012) bring these worlds to life for wide readerships.

Waste workers collect and transport wastes from point of origin, in garbage trucks or donkey carts. They sort and clean. Structured businesses in the formal sector include line supervisors, health and safety management officials, engineers and technicians, managers, executives, and staff. Waste laborers work in scrapyards and on landfills. They broker scrap transactions and trade. And workers and scrap dealers repurpose and sell things and materials that others have thrown away. They are also on the front line of immediate and long-term risks posed by wastes and waste disposal methods.

Waste work is critical to the functioning of modern cities, states, and countries (Wilson 2007). Vinay Gidwani terms this "infrastructural labor" (Gidwani 2015). Municipal trash collection should be a smooth, almost invisible service. However, when waste workers and authorities and/or corporations come into conflict, the results are tangible. Piles of garbage in the street can bring down governments, or at least weaken them. For example, garbage strikes in Naples in the early 2000s damaged that country's image, and highlighted corruption in government. Garbage worker strikes in the United Kingdom in the late 1970s are said to have turned public opinion toward Margaret Thatcher and the Conservative Party, who won the 1979 general election and ushered in a new era for the UK. A 2012 showdown with private garbage collection companies brought down the left-wing mayor of Bogota in 2013 when courts ruled against plans to incorporate informal sector organizations (he was later reinstated). The "rivers of trash" in and around Beirut in 2015, after the Lebanese government's efforts to draw up contracts with international corporations, generated massive protests and hurt the government badly. Protests in 2018 by Moscow residents about uncontrolled landfills and dumps roiled the city government and perhaps even the Russian leadership itself (Henry 2018).

Waste work cuts across scale and distance, from city streets to reprocessing and recycling facilities in distant countries. Small businesses in China melt down plastics imported from the US and Europe. Metals recovered from e-waste shipped from industrialized countries are smelted in Belgium. Engineers and technology specialists work on the latest innovations in waste-to-energy technology. Entrepreneurs raise money for circular economy startups that look to develop biological replacements for plastics. Corporate representatives and waste pickers alike have

sought voices in global climate and sustainable development policy at the highest levels. Waste workers, especially
informal sector waste-picking communities, have woken
up to the threats to their livelihoods, and seek to be recognized as the lifeblood of waste management in many
cities.

People who study economic production and labor across
the economy often differentiate between *formal* and *informal* labor and business sectors (Chen 2012). It is more
accurate to think of this distinction as a continuum, rather
than as demarcated sectors, and to note that individual
firms and workers may operate across both. The terms
are, however, a useful guide to the diverse ways the waste
work sector is organized around the world, also enabling
comparisons with other areas of work such as textiles, agriculture, hospitality and domestic service work.

Waste and recycling work are critical cases for studies
of formal and informal labor, mobilization and activism,
within cities and across borders. There is a human cost
of dealing with garbage but also the potential for livelihoods and upward mobility. However, in the coming years,
with privatization of municipal work and the emergence of
more forms of contingent labor (through the so-called "gig
economy," for example), we may see a "deformalization" of
waste work in developed countries. By contrast, informal
waste labor is being integrated into waste collection systems in many developing world cities.

The Formal Waste Economy

The formal sector consists of economic sectors and firms
where businesses and workers operate within official economic and legal systems. Businesses are incorporated
and pay taxes. Workers ideally have contracts and receive

social welfare and workplace protections. In the waste case they work for cities or in the private sector. Companies raise capital, through banks or the stock exchange, and are highly structured, usually hierarchical. In the waste sector, the largest companies are characterized by a global or regional reach, more advanced technology, and are often nested within larger utility – energy and water – providers.

The EU reported 1.443 million full-time jobs in the electricity, water and waste group within the environmental goods and services sector in 2015, up by over 50 percent since 2000 (European Environment Agency 2018). In the US, 2016 statistics show over 400,000 workers in waste management-related professions, with all but trash collectors earning more than the national median of $37,040. These are considered good jobs and sanitation workers are often unionized.[2] The largest corporations in the waste services sector employ tens of thousands across the world (see table 3.1), jobs that are highly differentiated by skill, function and pay.

However, industry safety is a concern. Sanitation work is classified by the US Bureau of Labor Statistics as the fifth most dangerous civilian occupation in the US: 31 sanitary workers died on the job in 2016. In New York City, while residential collection is done by city workers, businesses can choose their own carters, who pick up at night. An exposé from PBS's NewsHour in January 2018 investigated the nighttime world of safety violations, injuries, and deaths of pedestrians, cyclists, and waste workers in New York City ("Why Private Waste Management is one of the Nation's Most Hazardous Jobs"). Addressing safety conditions, through educating people in the industry and outside, is a top priority of the main industry organization in the US, the Solid Waste Association of North America (SWANA). The "Slow Down to Get Around" campaign is

Table 3.1: Major global waste management and resource recovery corporations						
	Waste Management	Veolia Environnement	Engie/Suez Environnement	Republic	Remondis	Covanta
Home country	USA	France	France	USA	Germany	USA
Countries of operation	USA, Canada	48	70	USA	Over 30, 4 continents	USA, Italy, UK, Ireland
Revenue*	$14.86bn (2017)	$27.8bn*	$22.25bn*	$9.4bn	$9bn	$1.7bn
Employees	42,300 (2017)	163,226	82,536	33,000	31,200	3,600

*Overall revenue, including non-waste business; 2016 data unless otherwise stated. Sources include: Statista.com and company websites and reports

targeted at motorists: being struck by a car is a leading cause of death for sanitation workers.

The waste management and resource recovery industry has been shaped by globalization, economic privatization, and urbanization. By the 2000s, a small number of multinational corporations based in the US and Europe dominated the global waste sector. These companies have, under changing corporate configurations, dominated the market for the past two decades and longer, performing waste collection, haulage, disposal and recycling services for cities, industry, and government agencies. In 2015, the solid waste industry worldwide was valued at $433 billion, a sum projected to rise in the years to come, particularly in South and East Asia, reaching $562 billion in 2020.[3] Industry actors have worked hard to transform the industry from its lowly origins to become one of the world's environmental services providers.

Waste Management Inc. is the largest waste management company in North America. In 2016, it ranked 549 in Forbes' Global 2000 list of top public companies in the world. It, along with Republic, controls over half the waste collection, transport, disposal, and recycling market in the US; its trucks are a familiar sight on the streets in US towns and cities. Clean Harbors provides hazardous waste management and spill and disaster remediation services. It started out with one truck and four employees in 1980. Another US company – Covanta – with revenues of $1.6 billion in 2015, and 5,000 employees – is one of the world's biggest waste-to-energy service providers, with operations in the US, Canada, and Ireland, and equity interests in Italy and China. Covanta processes 20 million tons of waste each year, enough to power one million homes.

European companies are often subsidiaries of larger utility providers – for whom water and energy provision are often much larger parts of their businesses. Suez Environnement's corporate family tree can be traced back to 1822; it took its name from its role in the construction of the Suez Canal in the 1860s. It provides waste and water services in 70 countries. Remondis is a German company that provides the full range of recycling services, including taking back packaging and electronic wastes as per EU directives. Veolia is the largest WTE provider in Europe. It holds contracts from Mexico City to Antarctica and is a leading player in WTE provision.

Despite relative continuity among the main players, the global waste management industry is a shifting terrain. The big firms acquire new ventures, divest themselves from others or from country markets. They merge with or take over smaller or regional players. Other smaller companies provide specialized services or serve regional markets. The industry also faces external pressures for change. Established companies and new entrants take on waste-to-energy, urban mining, and other activities, and retool to meet the challenges of the new global waste economy. Set against this global backdrop, local and city politics, and competition for contracts are still a big part of the waste business on the ground. For example, Spanish company FCC outbid Waste Management for a Houston recycling contract in January 2018 (even though WM's headquarters are there). On the ground, corporate greening efforts are not always successful. Emissions violations, accidents, dealing with cleaning up old sites and corruption still plague the sector, and these companies face significant obstacles in the US and elsewhere in siting new facilities.

China is now a global player in this industry. In 2015, Energy from Waste, Europe's market leader in WTE provision and valued at $1.6–2.1 billion was purchased by the Beijing Enterprises conglomerate. In 2014, Beijing

Capital bought New Zealand's largest waste management company for $800 million. Chinese companies have also recently invested in or purchased waste management and infrastructure companies across Europe, notably Poland and Spain. China's courtship of the European waste market is not surprising. It needs the technology and expertise to deal with its growing waste problem by building more advanced waste-to-energy and other waste facilities.

The global waste industry at the highest levels has actively distanced itself from the corruption and crime that has plagued waste management. To name just a few examples of a wide literature on the global North, this is chronicled in the US by Andrew Szasz (1986) and Alan Block and Frank Scarpitti (1985), and in Naples by Roberto Saviano (2008). It is premature to say the industry as a whole has shaken criminal behavior. Arrests for fraud and corruption happen among smaller players in the US. Criminal gangs and organized crime control waste work and resource extraction in many cities of the global South and also the trade in electronic wastes, given that many countries ban export and import (see chapter 4).

Finally, the mainstream waste industry faces significant challenges as it forges a role in circular economy initiatives, positioning itself as a provider of resource recovery and recycling services. It has a direct financial interest in maintaining waste diversion over waste prevention as well as access to the technology to enable effective recovery. These controversies have heated up as circular economy policies are designed and implemented (see chapter 7).

The Informal Waste Economy (Economies)

The informal sector consists, most generally, of businesses and workers who operate outside normal legal frameworks, such as tax, social security, and registration structures. Work is flexible, hours may be part time or well over a normal work week. It is precarious – it may dry up or be seasonal only, or workers can be fired at will, and is often (but not always) risky, carried out in unsafe conditions at home, in a yard, or a sweatshop.[4]

Informal labor exists across economic sectors in developing and developed economies – in agriculture, textiles, mechanics, household work, construction, transport, and many others, including waste and recycling. In 2009 (in the wake of the global financial crisis), the OECD found that 1.8 billion people – over half the world's workforce – were working without a formal contract and without social security (Linzner and Lange 2012, p. 69). Informal workers are likely to form the bulk of the workforce in highly hazardous enterprises – such as e-waste dismantling or ship-breaking – and under sweatshop conditions, for example in clothing manufacture. These are risks other countries have shipped overseas, the result of distancing (literal and figurative) in the global economy.

Informal sector workers, while they may operate outside state and market organizations and rules, are usually part of organized systems, whether hierarchy, collective, or extended family, with their own rules and systems of operation. They have global support. For example, Women in Informal Employment: Globalizing and Organizing (WIEGO) is a leading NGO in researching and advocating for informal workers.

Informal work is the longest-standing mode of organization in the waste and recycling, or scrap, sectors (Dias and Samson 2012). Estimates suggest approximately 20 million people worldwide work in the informal waste and recycling sector (ISWA 2012). A 2010 study in Brazil found over 229,000 informal waste workers working in that country alone (Dias 2011). In West and South Africa and in India, less than 1 percent of the urban workforce is engaged in waste work, but that still represents many people. In Zambia, UN-Habitat estimated around 60% of urban jobs are in the waste sector (UN-Habitat 2003).[5]

Waste collection and disposal services have only been formalized in developed countries for around a century, after considerable struggles, and remained an ongoing process for decades. David Pellow (2002) and Carl Zimring (2015) both document this history, and how ethnic and racial minorities, the main workers in the waste sector, had to mobilize and fight to gain the rights and protections of other formal sector workers. Dr Martin Luther King Jr. was in Memphis when he was assassinated on April 4, 1968, to support a strike by African American sanitary workers to protest working conditions and on-the-job fatalities.

The emergence of waste as a globally traded resource and commodity and the rise in awareness of how much waste is produced and disposed of around the world have shone a bright light on this informal sector. Informal waste picker organizations have made transboundary alliances that have placed their demands on the international stage and have collectivized at local levels to create more resilient, less exploitative networks and stable livelihoods. Their interests are represented by the Global Alliance of Waste Pickers (GlobalRec for short). The World Bank and other organizations have pointed to the informal waste sector as a way forward in developing waste management infrastructures in Africa, Asia, and Latin America's booming cities.

Delegates to the First World Conference of Waste Pickers in Bogota, Colombia, in 2008 adopted "waste picker" as their preferred term in the English language, a way of pushing out derogatory terms such as "scavenger" and enabling the development of a collective identity. In Egypt, they are the *zabaleen*, in Buenos Aires, *cartoneros*; *pepenadores* in Mexico, *catadores* in Brazil, *minadores* in Ecuador. In East and Central Europe, Roma workers do much of this work, and in India, it is done by Dalit, or members of scheduled castes, who are recognized as one of the most oppressed groups by the Indian constitution.

Waste pickers collect, sort, recycle, repurpose, and sell materials thrown away by others, extracting value from them. They go door-to-door collecting usable discarded goods for recycling or upcycling and reselling. Other informal waste laborers collect and dispose of household and other municipal solid waste left out for pickup. They clean and sweep streets and collect and sort the contents of waste in public bins and dumpsters. From collection to reprocessing, wastes go through several sets of hands, from street workers to small-scale local buyers, up to larger, regional traders and warehouse owners, creating hierarchies with collectors at the bottom (Gidwani 2015, p. 583; see also Beall 1997).

Many waste workers live on or right next to open waste dumps and landfills, picking out the still-useful, still-saleable items from the piles that have been discarded. ISWA's 2016 report on the world's 50 biggest open landfills shows that thousands of workers live on these sites (see figure 3.1).

In many developing country cities, the informal sector is the major or sole provider of waste collection and recycling services. One study cites Karachi, Pakistan, where 45 percent of total wastes generated are recycled by informal sector workers, none by the formal sector (Wilson et al. 2009). In Brazil, informal recycling accounts for high rates of cardboard (80 percent) and aluminum (92 percent) reuse (Dias 2016, p. 377). WIEGO and other organizations point out that waste pickers provide essential services to cities, at a far lower cost than alternatives: urban

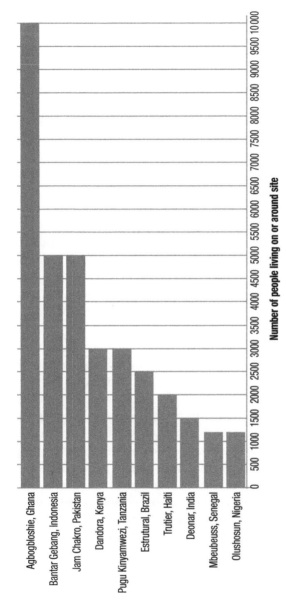

Figure 3.1: Number of people living and working on the world's ten largest waste dumps

Source: Data from ISWA 2016

cleanliness, beautification, increased waste diversion rates, and improved public health.

As a livelihood, waste picking and other forms of waste work can go beyond subsistence for the poorest or the homeless, especially with the opportunity to sell what they find and refurbish. In Mexico, Brazil, and Cairo (where informal waste workers are often in organized collectives or groups), workers earn several times the minimum (poverty line) wage (Linzner and Lange 2012, p. 78). One study found informal waste worker wages to be 110–240 percent above the local minimum wage (Scheinberg et al. 2010, p. 22). This is despite the low barriers to entry, in terms of skill, training, or needing a specific place of business.

Waste pickers, however, work in dangerous conditions, and are subject to accidents, disease, and death, in their workplace, on the road, and where they live. Over 750 deaths in dumpsites and landfills worldwide between December 2015 and June 2016 could be directly attributed to poor waste management (ISWA 2016). In March 2017, a landfill collapse in Addis Ababa killed at least 113 people (according to CNN, 38 males and 75 females, including children). One study found that waste pickers in Mexico have a life expectancy of 39, compared with 67 for the general population (Dauvergne and LeBaron 2013, p. 418), suggesting long-term health impacts that surpass the deaths by accident cited above. Wage statistics mask inequities, for example within communities that are strongly hierarchical and a minority of "bosses" and brokers benefit, while women and children earn less than men. Many children live and work on urban waste dumps around the world.

Waste picker communities suffer from exclusion, discrimination, and stigma surrounding their profession and livelihoods. Living in slums, shantytowns, and favelas they suffer from the lack of the very services they provide to wealthier neighborhoods. Even the richer among them face disrespect. They are often immigrants (from overseas or from rural districts). Normative perceptions remain that connect dirty places (settlements of waste pickers strewn with trash) with people who, therefore, are unhygienic, immoral, and possibly rebellious. They symbolize disorder. These views provide a rationale for authorities to displace these communities or replace them with formal enterprises, reinforcing historical perceptions and conflict based on ethnicity, class, race, or religion (McKee 2015). The Bedouin in Israel, Dalit in India, Roma in Bulgaria, and Kurdish refugees in Turkey suffer from such stigma (Dinler, 2016; Gregson et al. 2016; McKee 2015; Resnick 2015).

Sonia Dias, a leading expert on the informal waste sector, points out, "despite their visibility in most cities of the global South, waste pickers' contributions to the urban environment and economy are largely ignored by city planners who privilege a rational-modernist model of urbanization, based on the use of capital-intensive technologies imported from the global north" (Dias 2016, p. 377). Aspiring global cities make every effort to appear sparkling clean for foreign audiences, no matter what is done to remove dirt and the people officials see as visible symbols of poverty or as potentially unruly. All these motives were said to be in play in confrontations between waste workers and governments in places as diverse as Egypt, the Philippines, Argentina, and India. These and other conflicts between waste workers, the state, and the corporate sector are the topic of the next section.

Conflict on the Global Resource Frontier

Challenges to waste pickers' livelihoods, homes, and communities have come into sharp focus in the new global political economy of waste, in conflicts between global (formal) and local (informal) sectors in both North and South. The cases and examples in this chapter, along with others later in the book (see chapter 4), demonstrate how waste pickers are embedded in and across local and global networks and politics. Waste workers collect trash door to door in city neighborhoods, and not far away, others disassemble e-wastes shipped from the US or other countries thousands of miles away to the same city. Usually, these are not the same people or groups, but they find themselves united in common cause, dealing with incursions and competition on this globalized resource frontier.

Two common conflicts that pit governments and multinational corporations against waste workers are first, over who collects wastes, and second, over access to (or displacement from) garbage sites such as landfills and dumpsters, workers' source of livelihood. Processes of global neoliberalization, which have pushed governments to hollow out their powers and capacities in favor of the private sector and free-market globalization, have also shaped waste politics in cities around the world. Gidwani (2015) points out that ultimately MSW is usually the property of local governments, although they do not always choose to exercise their property rights.

Governments have arranged to work with private companies to provide waste services, often creating public-private partnerships. These may be local companies or multinational corporations. These strategies have been encouraged by international donors such as USAID, the UN Development Program and the World Bank (Gidwani 2015, p. 588). Where MSW collection has been contracted out to private entities, workers are typically paid less than what they earn working independently. Privatizing and enclosing landfills, for resource extraction (urban mining) or energy generation, displaces the people who live and work on them, quickly and without recourse or compensation.

In India, Egypt, South Africa, Mexico, Nicaragua, Colombia, the Dominican Republic, and other countries and jurisdictions, many of these conflicts are ongoing or recent, others date from the 1980s (see WIEGO and GlobalRec's webpages for these accounts). For example, in 2018, plans by Mexico City's governing authorities to build a waste incinerator that would take 4,500 tons of waste per day threatened to displace many of the city's informal waste workers (Sánchez 2018). A study of the enclosure of the La Chureca dumpsite, the largest in Managua, the capital of Nicaragua, and its conversion into a sanitary landfill, examines the impact on the 2,000 people who lived and worked there (Hartmann 2018). While they were moved into better housing, this survey shows their economic situation deteriorating in the absence of alternate jobs.

Strategies to exclude dumpster divers in wealthy cities, such as locking dumpsters, enacting trespass laws, and installing CCTV – as has happened in Vancouver (Wittmer and Parizeau 2016) – hide what city authorities call "disorder" but provide no remedy for underlying causes. The Freegan movement, which campaigns for the reclamation of discarded but edible food, has worked to resist these exclusionary practices to combat the discard of edible foodstuffs by restaurants and supermarkets in the struggle to end food waste (see chapter 5).

Waste picker organizations have also bonded across national borders to form transnational activist networks and found their voice on the international stage. The Global Alliance of Waste Pickers (GlobalRec), founded in 2009, connects thousands of groups from 28 countries, mostly in Latin America, Asia, and Africa. GAIA supports and partners with Zero Waste activists around the world.

Two cases illustrate the local and global connections in the new global waste economy: first between Cairo's waste pickers, the Zabaleen, and the Egyptian government, and second, transnational organization of waste pickers to highlight their interests at global climate change negotiations. Both cases illustrate the strength of activism, organization, and community ties, but also the ongoing nature of these conflicts.

Cairo's Garbage Collectors: The Zabaleen
Cairo, Egypt's capital city, has for centuries been one of the biggest cities in the Middle East. Like other rapidly expanding cities, it faces growing amounts of trash that formal services cannot handle. Since the 1940s, a close-knit community of garbage collectors, *Zabaleen* in Arabic, have performed those services, running an efficient and profitable service for the city.[6] There are 50,000–70,000 Zabaleen in the greater Cairo area, collecting around 12,000 tons of garbage per day. They live in their own villages within Cairo's boundaries and nearly 90 percent of the community are Coptic Christians, a marginalized minority in Egypt. The system they have put in place over generations is considered one of the world's most efficient resource recovery systems. They recycle up to 80 percent of what they collect. Yet their relationship with the state turned to confrontation in the 2000s. The Egyptian government, under pressure to follow dictates of international financial institutions to privatize, and its own goals of cleaning up Cairo to be more attractive to tourists and business, targeted the Zabaleen's well-established practices.

The Zabaleen – usually men and boys – collect trash and take it back to their villages, traditionally in donkey carts, where it is sorted by women and girls in the community. They contract with a group of middlemen – the

Wahiya – who have resided in Cairo for even longer. Both waste collection and pig farming support these communities. The Zabaleen separate out the recyclable material and traditionally they fed organic waste to pigs that were raised to be slaughtered and their meat sold, often to large tourist resorts in Egypt. The Zabaleen have invested significantly in training and technology to process waste more effectively. This success led to international support – including micro-lending programs and for community NGOs to work on safety and education. There is considerable income diversity among the Zabaleen. Many remain poor, some losing everything during the conflicts detailed below. Others are wealthy. They run reprocessing plants, sell products overseas, and live in Cairo's wealthy neighborhoods. Still, as with other recyclers around the world, their income depends on scrap prices, and on the political stability Egypt has recently lacked.

Over the years, the Egyptian state's relationship with the Zabaleen shifted from tacit acceptance to confrontation. Two events – Cairo's award of waste collection contracts to three European multinationals in 2003, and the government's decision to cull the Zabaleen pig population in 2009 – crystallized these conflicts. In the first instance, waste collection contracts with multinational companies cut the Zabaleen out of their traditional livelihoods – but also wound up undercutting Cairo's urban sanitation. As recycling rates fell and garbage went uncollected, residents preferred to bypass the new system and continued to pay their traditional waste collectors. In 2009, after global outbreaks of swine and avian influenzas, the Egyptian Ministry of Health announced that the Zabaleen's droves of pigs would be slaughtered – at first claiming that the move was to halt the spread of disease, then citing public sanitation and conditions in Zabaleen communities. Carried out as it was in the waning years of President Mubarak's rule, many suspected the move was made to appease Islamists, by further oppressing Coptic Christians. The pigs were slaughtered, the Zabaleen were unable to sell or make use of their carcasses and again, Cairo's trash began to pile up.

However, the Zabaleen's livelihood carried on. Cairo-based journalist Peter Hessler noted in 2014 that amid traffic, blackouts, water shortages, three constitutions, and three presidents, their trash was collected every day. This case demonstrates that communities and networks can be highly resilient in the face of a state that seeks to privat-ize (and globalize) urban services and space. The Zabaleen survived the pig cull, and authorities' attempts to "clean up Cairo" continue to meet with resistance. Even in April 2017, as local officials set up recycling kiosks in wealthy parts of the city, workers in Cairo's age-old informal recy-cling system remained combative.

Waste Pickers at Global Climate Change Conferences
As this account of the Zabaleen's relationship with Egyptian authorities illustrates, the conflicts informal waste work-ers are engaged in have globalized dimensions. Incentives for multinational corporations to reap larger-scale private benefits from taking over landfills, waste services, and rec-lamation activities in developing countries, often dovetail with government interests in mechanizing and formalizing landfill operations and removing waste workers from the street. This process of resource appropriation mirrors pop-ulation displacement from other large-scale development projects, such as large dams, that have displaced thousands. Chapter 4 also illustrates these trends, as multinational cor-porate interests – sometimes mining companies – move in to benefit directly from profits from urban mining and e-waste reprocessing.

The next case shows how localized waste work became connected with global climate politics, and to what extent such engagement worked to the movement's advantage, at least in terms of finding an international platform. Global climate negotiations have become an unexpected space for transnational waste picker activism, and a global venue for waste picker networks to advocate for their interests (Ciplet 2014; Newell and Bumpus 2012).

In 1997, the Kyoto Protocol, part of the UN Framework Convention on Climate Change (UNFCCC), established a series of mechanisms to facilitate greenhouse gas emissions reductions. One of these, the Clean Development Mechanism (CDM), was a global offset program. It was designed to enable developed countries to meet greenhouse gas emissions reduction targets by investing in projects in developing countries that met those goals at a lower cost, while also creating development benefits for the hosts. Typical projects included solar, hydro, and wind power installations. These projects generated emissions reduction credits that the rich country investors could sell on or apply to existing national obligations under Kyoto.

Waste-related projects made up close to an eighth of CDM-registered projects. Landfill gas capture projects are considered low-hanging fruit: landfill gas is mostly methane, a highly potent greenhouse gas. Stopping one ton of methane getting into the atmosphere is the same as stopping 72 tons of carbon dioxide in the short term. These projects are usually not costly to implement relative to alternatives, such as solar installation.

These projects became a focal point for waste picker activism and opposition in the late 2000s. Transnational waste picker and anti-incineration organizations started to attend conferences of the parties (COPs) to the UNFCCC and CDM meetings in 2009, up to and including COP

21 of the UNFCCC in 2015, the meeting that produced
the Paris Agreement. Activist networks have been able to
use these meetings as access points to protest the impacts
on their lives and livelihoods of landfill gas capture and
energy generation, and of waste-to-energy incineration,
and to demand participation. Waste-related projects drew
opposition because of waste picker and community displacement after landfill capping, and the danger posed
to surrounding communities from methane storage or
from incinerator emissions, ash, and other pollutants
(Vilella 2012). GAIA's reports specifically cite the Bisasar
landfill gas-to-energy project in Durban, South Africa,
the Timarpur-Okhla WTE project near Delhi, and the
Usina Verde incinerator in Rio de Janeiro as examples of
CDM-registered projects with negative impacts for environmental justice.

The Global Alliance of Waste Pickers and its allies
sought to raise awareness and concern about waste picker
livelihoods via these and other cases, recognition as a valid
constituency, and to include recycling-based projects in
global offset programs. They have protested peacefully outside convention halls, organized informative side events,
and lobbied for the inclusion of a new assessment method-
ology in the CDM that would allow local recycling efforts
to count toward emissions reduction credits. The fact that
waste picker organizations chose the climate governance
arena over global hazardous waste trade and chemicals
governance speaks to the reach of global climate politics
down into local communities. It also speaks to NGOs' skills
at identifying global strategies to highlight waste picker
concerns.

Episodes like this suggest that sometimes the international system opens to different actors in surprising ways,
even though the CDM itself failed in its goals. It also makes

visible the ways that climate solutions can harm the most marginalized populations.

Trends in Global Waste Work

In both North and South, waste work evolves and changes. This section investigates how structural changes are playing out, examining the growing integration of the informal waste sector into formal waste management in Southern cities and evidence of "deformalization" of waste work in the North. Formalization in the South is very different from its historical predecessors in the global North: trade unions as emerged in Northern cities are rare, and labor-intensive, low-cost work continues to dominate the introduction of capital-intensive technologies. In India, the government prefers to see waste workers as entrepreneurs (who bear their own risks) rather than as employees.

Organization and Integration of Informal Waste Work
Waste picker communities have had to respond to outside threats to their livelihoods, and in doing so, many have built lasting networks and strong relationships with each other, in the form of collectives or with local community-based organizations and environmental groups, as the above examples demonstrate. Waste-to-energy opposition campaigns have brought waste pickers together in common cause with home dwellers opposing neighborhood incineration projects. They have found allies in local authorities, churches, international NGOs, and international development organizations who favor micro-enterprises. Informal waste picking as a profession and a livelihood has, therefore, lent itself especially well to vibrant and strong activism.

The visibility of waste work may aid in these developments: outdoor workers in cities and towns can more easily reach each other or be reached by community organizations and journalists than are domestic, factory, or agricultural workers. However, the open nature of their work also leaves waste pickers vulnerable to harassment and violence, from other workers (for example conflicts over territory) or from authorities. Police are known to harass waste workers, often children or more vulnerable workers, in the Rio de Janeiro favelas. Such violence, however, has also galvanized waste workers to organize for self-protection. Gender inequity within the profession, worker safety and exploitation, and relationships with governments or outside corporations are part of movement agendas. Conflict, too, is visible: trash in the streets is a potent symbol of political disarray, and quickly galvanizes public opinion.

Efforts to organize work to protect workers in the sector, grant them representation, training, education for children, and basic health protections and benefits have gathered steam in the last two decades. Waste picker organizations or collectives may take the form of unions, cooperative organizations, or micro-enterprises (closer to small businesses). Waste worker collectives can be found all over the world. Waste pickers in Latin America started this process, in Brazil and Colombia, and waste workers around the world followed suit (Marello and Helwege 2018). In Guadalajara, Mexico, cooperatives, based on a network of families, disassemble e-wastes. Hasirudala in Bangalore works with companies to provide full waste collection services. Amelior, in France, is an example of a developed country collective. In South Africa, the Johannesburg Reclaimers Committee was formed in 2017 to resist the city government's efforts to shift recycling services to private sector companies (see below). The goal of many such informal organizations is to provide members with a sustainable livelihood: one that "can cope with and recover from stress and shocks, maintain or enhance its capabilities and assets, and provide sustainable livelihood for the next generation, and which contributes net benefits to other livelihoods" (Uddin and Gutberlet 2018, p. 2, citing Chambers and Conway 1991).

With the ongoing lack of formal municipal waste services in developing countries, the growing organization of informal waste workers, and the encouragement of international development agencies, more waste picker organizations have started working with city authorities and local industry to take over official waste management services as employees or contract workers, processes of inclusion, or integration. These city governments make an active choice to combine improving waste services without displacing the informal laborers who have held those jobs for decades.

Two particularly well-known examples of such integration are in Belo Horizonte, Brazil, and Pune, India (Dias 2016, 2017). Belo Horizonte is the sixth largest city in Brazil and was an early pioneer. Since the early 1990s, the city has worked with informal worker organizations, including them as preferential agents in the collection of recyclables in the city. In Pune, the Solid Waste Collection and Handling (SWaCH) cooperative started working officially with the city in 2008, responsible for door-to-door collection. They are far from a rare phenomenon, and international organizations like WIEGO work hard to create conditions for such integration to be successful around the world. In this they are supported by international organizations from the World Bank to the ISWA. Women waste workers have created their own organizations, to highlight their role – often as the workers least visible to the public, and subject to worse work and pay conditions even within broader collectives.

These processes of collectivization and integration are welcome. However, we know less about the conditions under which they succeed or fail. Governments change, competition emerges, either within the informal sector or from outside, and funding for training and capacity building might fail. Markets for recycled goods can crash. Collectives that reproduce pre-existing hierarchies are likely to perpetuate existing inequalities. Poor migrants are less likely to collectivize (or to receive support from existing community organizations), even though they are likely to do informal and dangerous waste work.

When interviewed by the media, waste pickers emphasize that they do not want their children to grow up in the same profession. Collectivization and integration mean that their children get more opportunities for education and care. At the same time, growing inequality and de-formalization at the level of the global economy may erode these opportunities. Child labor at informal waste work sites in China, Africa, and other parts of the world is an ongoing problem, as is conflict and violence against waste workers as state or corporate actors continue to enclose waste disposal sites. For example, GlobalRec reported that in Johannesburg in July 2018, waste pickers were attacked by members of a private security company specializing in evictions. This incident occurred within the context of recent city plans to implement a mandatory separation-at-source policy for households and contract out collection to private companies.

De-Formalization? Trends in Developed Countries
This chapter has charted trends toward integration – formalization – of waste and recycling work in countries in Latin America, Africa, and Asia. There is some evidence in developed countries, especially the US, that the reverse is

happening. There are more instances of contingent labor in the big cities, and anecdotal cases of informal recycling of e-waste and similar materials, adding to a shadow waste economy that has always existed (for example unauthorized bottle and can collectors). Protections for workers in wealthy economies have eroded, for example as trade unions are pushed back. Workers in the gig or sharing economy are part of the swelling "precariat," a term used to describe workers in contingent or part-time work (Friedman 2014).

Technological innovations, such as the automation of garbage collection and sorting, the use of drones to monitor landfill sites, and even driverless trucks, will shape employment trends throughout the industry (Rogoff and Spurlock 2017). The privatization of waste services and outsourcing scrap recycling overseas (to jurisdictions with cheaper labor) is driving a longer-term shift away from formal waste work as solid blue-collar jobs, with trained, paid (and often unionized) labor in developed countries.

European studies show that while municipal employment is still significant (and even edging its way back up in some cases), private companies dominate recycling and reprocessing. The EU's refugee crisis may have swelled the ranks of informal sector waste workers. Studies from the New York City sanitation department itself have found evidence of day laborers being hired to work regular shifts.[7] Certain types of waste and recycling workers are more likely to be excluded than others from the benefits of formal representation: private company employees or contracts, even if on a municipal contract, may not be paid according to local laws (Rosengren 2018a). MRFs in the US are nearly always privately run. They also have a higher rate of on-the-job accidents than waste collection work, according to the US Bureau of Labor statistics.

This trend toward deformalization, if occurring in countries like the US, is not all negative. Local collectives such as the award-winning Cero Co-op in Massachusetts, which provides compost pick-up around Boston, provide a locally based solution for specific cases. Likewise, small repair stores in Sweden provide a valuable local service and generate tax breaks. However, the question now is whether waste work will enter the "gig economy" in rich countries and cities? The "Uberization" of waste collection is a phrase tossed around, often without reflection upon what that might mean in practice.

Conclusions

This chapter has focused on how people derive their livelihoods from waste, from some of the largest global corporations to workers in the world's poorest urban areas. These actors work on a global resource frontier, even if their work is right outside the doors of their home. The frontier is a zone of conflict that engages governments, corporations, and workers in ongoing struggles, and uneven exposure to risk. The realm of waste and recycling work is a good place to see the confluence between the formal and informal sectors, and trends across and within the global North and South. It is also a central arena for new forms of transnational activism.

What about governance? This chapter showed forms emerging at the local level – collectivization, integration of informal workers – as well as the real challenges to governments in waste labor conflicts. The Waste Managements, Veolias, and Covantas of this world are high-profile corporations that can be held accountable more easily than less high-profile waste industry actors, but even so, they are routinely criticized for their records on the ground.

International organizations such as the International Labor Organization (Lundgren 2012) have spoken out for waste workers. Direct regulation of worker health and safety, wages and benefits remain in the hands of sovereign governments, however, and hard to tackle from the global level. Governments themselves may lack the will or capacity to address significant challenges in regulating transformed labor economies. Labor law and relations are changing, for better or for worse, and the status quo is no longer fit for today's realities (Ashiagbor 2019). This chapter has shown innovative ways waste workers have mobilized to get their concerns heard in a global economy. Chapter 4, on resource extraction from electronic wastes, continues these themes.

CHAPTER FOUR

Discarded Electronics

In 2002, the Basel Action Network (BAN), the leading NGO campaigning against the global hazardous waste trade, published a ground-breaking report *Exporting Harm: The High-Tech Trashing of Asia* (Puckett et al. 2002; see also Puckett et al. 2005). It exposed how electronic wastes, including old lap-tops, cellphones, TVs, stereos, and other discarded equipment are shipped from Europe and North America to the global South. The report showed how workers dismantled old electronics under highly hazardous conditions to extract gold, copper, and other valuable metals for reprocessing and resale. Many of these pieces of old equipment (including my very first Apple Macintosh desktop computer, bought in 1991) were shipped to developing countries as charity – to "close the digital gap" between North and South. Instead, they found their way into junkyards in Africa and Asia where badly-paid laborers, including children, wash down old circuit boards with acid and burn old computer casings, exposing workers to dioxins, mercury, lead, and other toxic substances. The town of Guiyu in Guangdong Province, China, supported, at its peak, over 5,000 workshops, with workers disassembling 15,000 tons of e-waste each day. Much of this waste came from the US and Europe.

This report, and others published around that time (including an hour-long exposé on the US news show *60 Minutes* in 2008) hit a nerve with the public. Until the early 2000s, there had been little critical examination of the new, "greener" high-tech industries' environmental and labor records. Also, e-wastes have a personal connection: the phones, tablets, and laptops that we replace on a routine basis are still recognizably those products when they show up in junkyards thousands of miles away.

National and international authorities immediately started taking measures to reduce or halt these transboundary flows of e-wastes. Well into the second decade of the twenty-first century, however, the e-waste trade is still going strong, and the basic assumptions that made the global trade in e-waste and used electronics a defining global environmental justice issue of our time no longer hold.

The sheer size and global diffusion of e-waste stocks, their mobility, and the entry of global capital make e-wastes the epitome of a global resource frontier. Old devices and appliances can be dismantled and reprocessed and metals extracted for valuable metals and other materials. They can be repaired or refurbished and resold. However, they also come with significant risks, for the workers who disassemble or repair them and for the health of local communities and environments, even as they provide a livelihood for hundreds of thousands (perhaps millions) of workers, mostly in the informal sector – around the world.

The trade has changed over the years. The e-waste trade is no longer just a "North-to-South" problem. Many discarded electronics are shipped between developing (non-OECD) countries, or from poorer countries to richer (Lepawsky 2018). Many middle-income and poorer countries now produce substantial amounts of e-wastes themselves. Ghana imports used electronics from 147 countries (Grant and Oteng-Ababio 2016, p. 9). E-waste from the Caribbean is shipped to Venezuela, while Middle Eastern countries ship to South Korea or Indonesia. Multinational mining corporations now engage in harvesting "urban ore" and extracting resources, rebranding themselves as sustainable in the process.

Two emerging narratives of the e-waste trade, or, what many call the trade in used electronics, have created debate and contention. The first narrative contests the 'rich country perpetrator'–'poor country victim' narrative, and highlights complexity and changing global political economies. The second contests the depiction of e-waste sites that has dominated the media – that conditions for workers are unavoidably horrific, most of what is shipped is dismantled under dangerous conditions, and most of that is unusable. Instead, some argue, much e-waste is repairable, and in fact is repaired and sold on in local or foreign markets, one reason patterns and trends in global e-waste shipments have confounded the Basel Convention and other national and global regulatory efforts.

The new global political economy of e-waste challenges traditional governance of production flows and disposal of waste. Yet, it has also ushered in governance innovations, such as extended producer responsibility (EPR) campaigns implemented in the EU and California. This case brings home the importance of understanding the entire lifecycle of a product, from design to final disassembly, and within that lifecycle, all the points along products' supply chains. Many of these governance innovations come back to product design, to maximize reparability and minimize toxicity. As with other cases in this book and elsewhere, the most effective solutions are those that enable the extraction of resources or value from discarded materials while minimizing risks.

Electronic Wastes: What, Why, How Much?

The rise of e-waste as a policy issue in the 2000s initiated efforts to define and quantify the problem. Discarded – or waste, electronic and electrical equipment (WEEE), to use the EU's official terminology – makes up one of the fastest growing and most complex waste streams. According to the 2017 *Global E-Waste Monitor*, considered the most authoritative source on e-waste data, countries collectively produced 44.7 million tons of e-waste in 2016, or 6.1 kg per inhabitant of the planet, statistics that are likely underreported (Baldé et al. 2017). This number compares with 41.4 million tons in 2014 and a projected total of 52.2 million tons in 2021, with Asia the largest overall source.

So far, only 20 percent of the e-waste produced worldwide is documented as collected and recycled under formal take-back programs (Baldé et al., 2017), by producers, municipal authorities, or collection centers. The rest is undocumented (much is effectively hoarded in basements, attics, and offices) or diverted to landfill or incineration. These data overlook informal recycling industries in developing countries, where rates are far higher. In Ghana the collection rate of used devices is 95 percent (Grant and Oteng-Ababio 2016, p. 6).

E-wastes consist of "almost any household or business item with circuitry or electrical components with power or battery supply" (Baldé et al. 2017, p. 11) that has been discarded by owners or users. They vary as to where they came from (households or businesses), what they are, what state they are in, and what they contain, both valuable and hazardous.

Electronic wastes include personal electronics, such as cell phones, and large appliances, from dishwashers to cars. There are six main categories (Baldé et al. 2017): temperature exchange devices (fridges, air conditioners, freezers); screens and monitors (TVs, laptops, tablets); lighting; large equipment (washing machines, stoves, copy machines, solar panels, servers and data storage units);

small equipment (vacuum cleaners, microwaves, toasters, tools, toys, medical devices); and small IT and telecommunications equipment (phones, GPS devices, routers, hard drives). In fact, personal electronic devices make up only a small fraction of the e-waste stream by volume (see figure 4.1). Appliances old and new continue to make up a sizeable proportion of e-waste stocks. Older, bulkier items, such as cathode ray TVs or traditional refrigerators continue to make up a sizeable proportion of e-waste stocks, although fewer enter the waste stream these days. On the other hand, newer appliances now contain processors, circuit boards, monitors and other equipment that have blurred the line between electronic and non-electronic equipment.

E-waste generation is driven by the growth in the size and diversity of the electronics and information and communications technology (ICT) sectors worldwide. *Time Magazine* reported in March 2013 that more people now have access to cell phones than to adequate sanitation. Owning multiple devices is commonplace, along with their chargers and batteries: the average household in the US has 28 devices, according to a 2013 Consumer Electronics Association Report. The *2017 Global E-Waste Monitor* reports a global total of 7.7 billion cellphone subscriptions (Baldé et al. 2017).

The spread of computerization is not limited to new devices. Vehicles and equipment – from cars to agricultural and industrial machinery – have been remodeled and redesigned to incorporate computing technology. With half the world's population online, the hardware infrastructure has grown to meet demand. Google alone has well over a million servers to run its search engine and other operations. Lithium batteries in electric cars and solar panels, vital components of green growth, also end up as e-waste (Winslow et al. 2018).

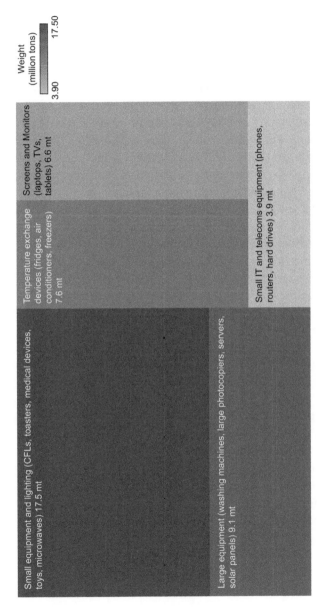

Figure 4.1: E-waste generated globally by type and weight

Source: 2017 Global E-Waste Monitor (Baldé et al. 2017)

Stocks of older equipment, including cathode ray TVs and computer monitors, are bulkier than the most modern consumer electronics, and still make up a large part (by weight and volume) of the e-waste stream (see figure 4.1). They also contain materials that are dangerous or toxic if mishandled, such as plastic, chemicals, and leaded glass. Old refrigerators and air conditioning units taken out of circulation still contain chlorofluorocarbons (CFCs) and hydrofluorocarbons (HCFCs), both of which are ozone layer depleting substances and have an extremely high global warming potential if released into the atmosphere.

Third, e-waste accumulation is driven by planned, or built-in, obsolescence, when companies design their products to have an artificially limited lifespan. This is not a new concept – one of the earliest incidences of planned obsolescence at an industrial scale happened in 1924 when representatives from the global lighting industry agreed to a 1,000 hour standard for the lifespan of a light bulb, down from a level that had alarmed the industry and threatened sales.[1] General Motors introduced the "model year" for cars at about the same time, to induce customers to switch up more frequently.

We see this trend today. Companies and advertisers fan consumers' desire for the next great device, version or upgrade. Our devices and appliances are no longer designed to last for years or be easily repaired – a point we will come back to. The laptop I am currently typing on has a battery and hard drive purposely sealed in by patented screws, making it next to impossible to replace. Smartphones are used on average for 21.6 months in the US, 19.5 in China, 18.8 in Germany (Baldé et al. 2017, p. 21). The fact that much of what we use – not just electronics – is designed to fail over a distinct timeframe is part of our culture of disposability (Acaroglu 2018).

Box 4.1: Urban ores
Ferrous metals: Iron and Steel
Non-ferrous metals: Aluminum, Copper, Lead, Tin, Titanium, Zinc and alloys such as Brass and Bronze
Precious metals: Gold, Silver, Platinum, Palladium
Rare metals: Cadmium, Cobalt, Indium, Lithium, Mercury, Niobium, Tantalum, Tungsten, Zirconium
Rare earth elements: Cerium, Europium, Gadolinium, Lanthanum, Terbium, Ytterbium
Building materials: Concrete, cement, bricks, timber
Other usable materials:,Plastics, glass, rubber

E-Wastes' Promise: Extractive Value . . .

For the 2020 Tokyo Olympic Games, the Japanese government pledged to make all the gold medals from recycled gold. This gesture recognizes the wealth of materials that lie in "reservoirs" in and around human settlements around the world. These could be iron and steel pipes and construction debris, or they could be the variegated collection of metals that go into our modern-day electronics.

The *2017 Global E-Waste Monitor* (Baldé et al. 2017) estimates the total value of all raw materials present in e-waste at approximately €55 billion in 2016. They include precious metals (gold, silver, copper, platinum and palladium), ferrous and non-ferrous metals (iron and aluminum), and recyclable plastics (see box 4.1). Further statistics include:

- In 2014, the roughly 300 tons of gold extracted from e-waste added up to around 11 percent of global gold production from mines (Baldé et al. 2015).
- For every million cell phones we recycle, 35,000 pounds of copper, 772 pounds of silver, 75 pounds of gold and 33 pounds of palladium can be recovered, according to

the US EPA's Electronics Donation and Recycling web-page.[2]

- Globally the amount of copper in such waste deposits [reservoirs] is of a similar magnitude (390 million tons) as the current in-use stock of copper (350 million tons) (Krook and Bas 2013, p. 4), while around 34 percent of all global steel production comes from recycled material (Arora et al. 2017, p. 215).
- E-wastes also contain rare metals, including lithium and cobalt, and rare earth elements that are needed for batteries used in hybrid and electric cars. Demand is growing, but there are no substitutes (Zhang et al. 2017). They are found in small, scattered deposits, and are therefore hard to mine.

These secondary stocks of extractive resources are collectively called "urban ore" because they are often located in or shipped to urban areas and their peripheries. Reservoirs of secondary ores exist all over the world, wherever there is a scrapyard, landfill, dumping ground, construction or demolition site, even attic or basement. Therefore, urban mining is the process of extracting or reclaiming valuable resources from existing reservoirs of discarded materials: end-of-life vehicles, construction and demolition sites, and other repositories within the 'technosphere' are rich sources of value, including steel girders, copper piping, bricks and concrete, and discarded electronics (Krook and Baas 2013).[3]

Mining is most commonly associated with the extraction of primary minerals, coal, precious metals, and gems from below-ground repositories. Territorial mines are fixed in place, dug deep into the ground (or sea bed) in the quest for uncertain deposits of ore. They are often located in mountain or forest regions far from large human settlements. Urban ore is richer, of a more predictable quality, and more accessible. The technologies used to extract it are quite different from the drills and digging equipment associated with traditional mines, but the equipment for smelting and refining the ores can be the same.

Urban mining substitutes for extraction of scarce primary resources. It can replace mining of metals and precious stones in war zones, the "blood" metals and gems that are used by military and paramilitary organizations to fund ongoing conflicts. It avoids greenhouse gas emissions, and limits environmental costs associated with traditional mining practices, including large-scale ecosystem destruction and land degradation, and large-scale community displacement. Globally, below-ground mining has one of the worst fatality rates of any occupation. According to the International Labor Organization (ILO), mining accounts for 1 percent of the world's workforce, but 8 percent of workplace deaths.

The term "flexible mine," introduced by Freyja Knapp (2016), captures the characteristics of above-ground, or urban, mining that have encouraged the trade, and drawn traditional mining interests to invest in the field. It avoids the sunk costs in time and capital that traditional mining incurs through exploration, site construction, and risk management (including permitting and environmental impact assessments). These mining sites are easily moved (in that equipment can be packed up and relocated) and urban ore can be shipped to the same smelting facilities that process virgin ores. From mining companies' perspective, shifting at least part of their operations to this sector, and harnessing themselves to greener practices of recycling and reprocessing, counterbalances the industry's reputation for damaging exploitation.

Agbogbloshie is in an industrial district in Ghana's capital, Accra. A former wetland, it had for years been a destination for automotive scrap, and now is one of the world's main destinations for e-waste, from Ghana and up to 147 other countries. It is the archetype of an urban mine. At 10.6 hectares, it is one of the biggest in the world, and around 40,000 people live and work in the immediate area. It houses both repair shops and dump areas. In the former, devices are repaired and sold on. In the latter, valuable ore is extracted and often exported. What is left is often burned, and the area is significantly polluted, with workers' health severely affected. The networks of trade and of people who run import, resource extraction and repair, and export operations are detailed below.

. . . and its Perils: Magnified Risks

> For money, people have made a mess of this good farm-ing village. After they have dismantled the computers, they burn the useless parts. Every day villagers inhale this dirty air; their bodies have become weak. Many people have developed respiratory and skin problems. Some people wash vegetables and dishes with the polluted water, and they get stomach sickness. – Mr. Li who has lived in Huamei village for 60 years – one of the villages that makes up Guiyu, quoted in Puckett et al. (2002), p. 15

The risks of e-waste disposal and recycling are magnified by the growing quantities of e-waste generated around the world, and by its displacement to parts of the world with laxer regulations and fewer safety measures. The trade rep-resents distancing of risk by rich country consumers and businesses. For them, the worst impacts fall far from where benefits are enjoyed. This, incidentally, is true of both the production and disposal of most electronic goods.

Dealing with discarded electronics poses serious health and safety risks to the workers who dismantle it, to local communities, and to local environments (Grant et al. 2013). These risks are magnified as quantities of e-wastes increase, and as they flow to parts of the world – in developing and developed countries – less equipped to deal with them safely. Ongoing reports from BAN and the Silicon Valley Toxics Coalition (Puckett et al. 2002 and 2005, Basel Action Network, 2016), the International Labor Organization (Lundgren 2012), and the World Health Organization's E-waste and Child Health Initiative show vividly how dangerous e-waste dismantling can be. That these reports have appeared over the course of so many years attests to the ongoing and serious nature of the problem.

E-wastes contain mercury, lead, arsenic, cadmium, and other dangerous substances that can be released when the wastes are treated under the wrong conditions. They emit noxious chemicals when burned, or fumes when acid is used to strip out valuable elements. Toxic run-off leaches into the ground. Compact fluorescent lightbulbs, flat screens, and laptop screens contain mercury that is easily released when they break. When plastic computer casings are burned, they emit dioxins, a persistent organic pollutant. Toner cartridges emit microparticles of dust into the air. Shattered glass from cathode ray tubes (CRTs) is an immediate hazard, and CRT glass also contains lead. CRT glass is often just buried. This is a problem just over the US border with Mexico, areas where the ground is clogged with shards of glass.

E-waste disassembly lends itself to informal sector businesses and labor. Dismantling sites exist in city neighborhoods and villages around the world, often coastal areas. Some are better than others. Guiyu and Agbogbloshie have received a lot of international attention and have responded to it, especially in China, where Beijing cracked down on informal worksites in Guiyu in the mid 2010s.

The Basel Action Network's Global Transparency Project (2016) has investigated illegal exports from the US, tracking discarded electronics to China, Hong Kong, Thailand, the

Dominican Republic, and Pakistan, finding legal violations and dangerous public health and safety conditions. Much e-waste disassembly is done by hand, which can be difficult and dangerous. Workers have been photographed using acid to strip circuit boards down to their valuable elements. They often have no safety equipment other than a paper mask, and a fan to clear fumes. Child labor is prevalent (Lundgren 2012, p. 20).

Workers in this sector frequently earn above subsistence wages, and there are low barriers to entry, in terms of skills and set-up costs, into this workforce (see chapter 3). Sometimes e-waste disassembly is the only work available in a community, all factors that outweigh (at least in the short term) wider health and environmental costs. Studies of the immediate and long-term impacts of e-waste, particularly on high-risk groups, around Guiyu found disturbing results: elevated lead levels in children's blood, a higher incidence of birth defects, respiratory illness, low height and weight, and other evidence of high burdens of chemicals in their bodies (Grant et al. 2013).

Workers are also vulnerable to conflicts associated with the resource frontier. Not only is it dangerous work in and of itself, and insecure (as with most informal labor), workers are prone to conflict among themselves or between groups over access to repositories. They are also vulnerable to exploitation by their bosses or by lack of official oversight. Organized crime groups are still involved with e-waste transport and disposal (Gibbs et al. 2011), especially given the illegality of export and import in many parts of the world.

Another concern about e-waste reclamation is data security, especially for the original owners of the device, be they government, corporations, or individuals. Sensitive or proprietary data can be, and frequently is, harvested from discarded computers and other devices.

Does Urban Mining Achieve Its Potential?

Finally, given the *potential* value that can be realized from e-waste recycling, to what extent is that value realized? E-waste collection and processing can be highly inefficient, with low recovery rates overall (Zhang et al. 2017). First, collection of e-wastes for recycling is challenging. Many simply store or warehouse old devices (Saphores et al. 2009). Curbside recycling services usually will not take e-waste, leaving users without easy options other than land-fill. Second, available techniques for separating urban ore from discarded devices and appliances are not necessarily efficient, whether advanced pyrometallurgical recovery or separation by hand. High-tech equipment is safer for workers and extracts more value, but it is expensive to buy and operate. Third, the market for secondary resources is volatile, and dependent on prices for primary equivalents. The price of scrap is always lower, although the difference varies over time. These prices determine how much effort is put into extracting and reprocessing urban ore (Jolly 2007). The scrap market also has fewer mechanisms for hedging risk than the primary one (Xiarchos and Fletcher 2009).

It is also the case that the price of a refurbished second-hand laptop or phone on the open market is much higher than the value of the metals contained in them, at least in developed countries. A study cited in Baldé et al. (2017, p. 54) finds that the metals in the average cellphone are worth €2, far less than prices for second-hand phones. We return to this point later.

On the Global Resource Frontier:
The International Trade in E-Wastes

Discarded electronics and electrical equipment are found and recycled or reprocessed all over the world; they are highly mobile, globally dispersed, and contain considerable extractive resources with high market value. As such they exemplify a newly recognized global resource frontier. This resource frontier is attracting large-scale capital and small businesses, but it is also characterized by conflicts among producers, dismantlers, and government agencies, with elevated levels of risk of exposure to toxics often shouldered by informal sector workers.

The e-waste resource frontier is fundamentally shaped by the trade in electronic wastes: the shipment of discarded and used electronic devices and appliances from one country for reuse, repair, or reprocessing in another. The trade demonstrates the mobility of this resource frontier as electronic wastes move from point of discard to point of extraction and disposal – and on to new destinations. They sometimes wind up back where they started, the parts of the world where the biggest electronics manufacturing plants are located, where they can reenter global circulation as parts of new products (Iles 2004).

Two narratives complicate easy assumptions about the e-waste trade. First, trade routes are more complex than assumed early on. E-wastes do not all travel on a linear path starting in Europe or the US and ending at their destinations in Africa or Asia. Second, not all or even most e-waste is stripped of value and the remainder dumped on its arrival. More is repaired than many think. These narratives undermine a Southern victim narrative and draws attention to the global networks needed for recycling to happen. In terms of governance, there are no simple solutions. A North–South ban, a popular option, would be obsolete quickly. This does not negate responsibility of Northern country producers and consumers, and nor does it mean risks will be significantly reduced without regulatory action. However, it does fundamentally alter the governance landscape in this highly complex arena.

The e-waste trade has two antecedents: the global trade in hazardous waste and the global scrap trade. Both these trades will come up again in chapter 6 on the plastics trade.

The Hazardous Waste Trade

The global waste trade is not new. Its most familiar form is the global trade in hazardous wastes that emerged in the 1980s, with egregious cases of toxic wastes from OECD countries dumped in Africa, the Caribbean, Eastern Europe, or simply in the oceans, to avoid local regulations and costs. Even then, the trade was not that simple. Much of it was legal, with wastes shipped among OECD countries (O'Neill 1997, 2000). Many proposed shipments to developing countries were turned down by their government authorities, but the problem of waste dumping gained political traction in the global arena, sparking the creation of a new international environmental agreement.

The 1989 Basel Convention on the Control of Transboundary Movements of Hazardous Wastes and Their Disposal, sponsored by the UN Environment Programme, first sought to control, and then to ban North–South waste dumping but ran up against arguments from many non-OECD countries that such a ban robbed them of their ability to recycle or reprocess imported wastes for their valuable materials, regardless of their ability to deal with them in safe or environmentally sound ways.

The global e-waste trade looks quite different from the trade in hazardous wastes that raised so much concern and galvanized global political action in the 1980s. These were primarily industrial wastes: ash, sludge, old batteries, often with little recycling potential. To take a more recent example, in 2006 truckloads of toxic sludge shipped originally from Europe were dumped on open landfills in Abidjan, the capital of Ivory Coast, causing several fatalities. Compare this shipment with a typical shipment of electronic wastes made up of whole or broken appliances and devices, which look different (recognizable as things), and are treated very differently, but still have deleterious effects on workers, communities, and environmental health.

The Scrap Trade
A significant global trade in scrap of all kinds has carried on outside the purview of global environmental governance, but is a sizable force in the global economy. This trade includes problematic types of scrap such as electronics (as discussed in this chapter) and plastic (see chapter 6). According to the Institute of Scrap Recycling Industries (ISRI), 800 million tons of scrap is consumed annually worldwide: it is often the first link of the manufacturing supply chain. The scrap industry is worth more than $105 billion annually in the US alone, and in 2017, ISRI reported 155,632 jobs in the US scrap sector, with average wages and benefits of over $75,000. Scrap companies (large and small) collect, process, broker, sell, and recycle post-producer and post-consumer metals, paper, plastics, glass, rubber, electronics, and textiles.[4]

Since the earliest recorded instances of cross-border scrap shipments, including brick dust sent to Moscow from London in the early nineteenth century (Wilson 2007, p. 199), the scrap trade has grown into a huge global business. Its reach has been facilitated by economic globalization, which led to more trade, lower import tariffs, and lower transportation costs. The most visible forms of scrap trade have been those that are problematic: electronic wastes, and now plastics, but the bulk of the trade has been in steel, aluminum, copper, and other relatively non-problematic scrap.

Scrap imports and exports can be tracked through UN Comtrade's database.[5] According to the ISRI Scrap Trade Database in 2016, global exports of ferrous, paper, non-ferrous, plastic, rubber, and other scraps totaled nearly 150 million metric tons, with an estimated value of $85.25 trillion. China and the US have been the largest players. However, other countries lead the field. Turkey, for instance, was the largest importer of ferrous scrap in 2016. India imports large quantities of metals such as aluminum and zinc (but not, historically, plastic or paper scrap). Scrap tends to go to countries where it is most likely to be reused in manufacturing, not the ones with the cheapest labor and disposal costs (Minter 2013).

In his book about the global scrap metal industry, *Junkyard Planet: Travels in the Billion Dollar Scrap Trade* (2013), Adam Minter recounts the scrapyard and dealer networks that are the lifeblood of the scrap industry, as well as how market volatility shapes the fortune of those who run it. He witnessed firsthand the impacts of the 2008 global financial crisis on scrap markets in the US and China, when prices of primary metals plunged, and scrap prices even further, by as much as 80 percent. Importers in China simply walked away from container loads of scrap and the deposit they paid, rather than take the loss. It took years to clear the backlog, make back the losses and rebuild trust between sellers and buyers. No matter how bad the 2018 recycling crisis caused by China's halt to certain scrap imports, it is "nothing compared to the ugliness that followed Lehman," referring to the 2008 collapse of the financial conglomerate that marked the beginning of the Great Recession of the 2010s.[6]

How Much E-Waste Crosses Borders?

In BAN's report, *The High-Tech Trashing of Asia* (Puckett et al. 2002), one statistic reported in the Executive Summary sticks out: "informed industry sources estimate that 50–80% of e-waste collected for recycling in the US is not recycled domestically at all, but very quickly placed on container ships bound for destinations like China" (Puckett et al. 2002, p. 4). This number is cited throughout studies and news reports even today, even though it is based on a single informal source in the report, and the authors qualify its veracity when scaled up (Lepawsky 2018, pp. 33–5).

It is impossible to determine how much e-waste is shipped across national borders with any level of accuracy. Despite the visibility of the shipped electronics, the workshops and yards, and the workers themselves, the actual trafficking in e-waste is nearly always illegal. It contravenes international law or laws of importing or exporting countries, further complicating efforts to track routes, points of export, and destination. We see the e-wastes piling up in poor communities and countries, but we lack the ability to see exactly from where they originate.

One way to estimate such data is through meta-analysis. Meta-analyses pull together existing studies to come up with comparable ranges of data and outcomes. One of these, from 2014, puts annual exports of e-wastes to non-OECD countries at 5,023 kilotons (kt) (with an uncertainty range of 3,642–7,331 kt), representing 14.4% (10.4–20.9%) of global e-waste generated, or 23% (16.7–33.5%) of e-wastes generated with the OECD (Breivik et al. 2014).

The digital era itself has enabled innovative ways to estimate these quantities. The Basel Action Network attached GPS trackers to 205 discarded devices between 2014 and 2016. Of those, 69 were confirmed as exported. Most of those wound up in Hong Kong, mainland China, or Taiwan. The Reassembling Rubbish project, headed by geographer Josh Lepawsky, used data from the UN Comtrade database (which compiles reported global trade across goods and services). The project researchers found that between 1996 and 2012, total global e-waste flows more than doubled (noting that this data covers only reported trade).

The Solving the E-Waste Problem (StEP) Initiative is a multi-stakeholder initiative coordinated by the UN University (see below). Its "Person in the Port" project in Nigeria had an investigator inspect containerloads of goods. In their report, published in 2018, they noted finding e-wastes smuggled in containers, much of which came from the EU. There is a thriving used car trade from Europe and Asia into Africa, and many of the used electronics they found were packed into these cars.

These examples show how difficult it is to obtain reliable data on e-waste shipments across borders, although they also demonstrate ingenuity in quantifying a significant global issue. It is better to have incomplete rather than bad data to make good policy; we must be careful about using poorly based data without significant qualifications. The exact scale of the e-waste trade is unknown. However, we know there are large flows of e-wastes across national borders, and we are beginning to better estimate their routes and directions.

Trade Routes

Export of discarded electronics from many OECD countries has historically been relatively easy and has been underway since the mid-1990s at least. Users drop their items off with a retailer or collection operation, they are picked up, sold to, or bought by brokers, shipped to ports in Africa and Asia, where they are then transported to dismantling and repair yards. Exporting is cheaper than dealing with the wastes at home. Even if restrictions are in place, shipments can be small and fly under the radar, especially as their contents are stable and safe until they reach point of disassembly. The EU was the first to crack down, so in theory export from Europe is harder, but still occurs.

In May 2018, police in Thailand displayed seven shipping containers each packed with about 22 tons of discarded electronics, imported by companies in Thailand without permits (Reuters, May 30, 2018). Amid concerns about becoming the e-waste dumping ground of the world after China closed its borders to e-waste in 2015, the Thai government had declared it would also ban these imports. Officials noted these e-wastes came from all over – such as Japan, Hong Kong, and Singapore, for example – and were impossible to trace. Hong Kong is a transit state for many types of waste and scrap.

The prevailing picture of the e-waste trade starts to change once one looks at the demand for imports. Importers of e-waste pay to bring in used electronics. The market is not solely driven by suppliers paying brokers to take them away. Recent studies depict more complex flows and networks of importers, exporters, disassembly work, and markets of and for e-wastes. Lepawsky and colleagues have compiled databases and visualizations of e-waste flows that track complex webs of transfrontier movements (reassemblingrubbish.xyz). Between 1996 and 2012, they found a substantial drop in the overall proportion of North–South e-waste flows, and growing evidence that Southern countries were exporting e-wastes, often to countries with much higher GDP.

In fact, this and other analyses paint a challenging but fascinating picture. E-wastes flow from China to Nigeria, for example, or from the Middle East to Indonesia and Korea (Lepawsky and McNabb 2009). India has imported e-waste from African countries. China has, just in the 2010s, cut its imports back, and as of 2015, little imported waste is getting into Guiyu. US exports have been redirected to Mexico. However, e-waste sent to the Chinese market often winds up in Hong Kong, whose special economic status places it outside mainland China's borders for trade purposes. There it stays or can be quietly shipped across to the mainland (Basel Action Network 2016). In Asia, the biggest importers are Korea and Japan, both high- and middle-income countries. Finally, domestic production of e-waste across developing countries is ballooning as digital technology catches hold. China has more than enough of its own to process now. We now know that discarded electronics are often re-exported to other countries in scrap or refurbished forms. They are not just for local markets in China, Ghana, or Nigeria, nor are they dumped. In other words, e-wastes' journeys are far from over when they reach their site of reprocessing. Studies of e-waste import, processing, and re-export in Ghana by Grant and Oteng-Ababio (2012, 2016) capture this complexity. Imports come from Europe (often as charity goods to avoid export bans), the US, but also China, Hong Kong, South Africa, and the Gulf States. As Hong Kong, Durban, Mombasa, and Dubai are known transit states for e-waste, its origins are murky. These countries have exported scrap to China, India, Europe, Australia, and many other countries (Grant and Oteng-Ababio 2016, pp. 6–14). China re-exports repaired/refurbished electronics to Southeast Asian countries (Lundgren 2012, p. 150).

Thus, the shape of e-waste trade is not determined simply by where the cheapest labor and lowest environmental regulations are. E-waste mostly goes to where it can be used and recycled, and where relationships between brokers and importers exist.

Networks and Relationships

Discovering who controls shipping, export, and import decisions and runs dismantling enterprises – in other words, who benefits from e-waste trading – is difficult. This resource frontier is a competitive, even conflictual zone, as one would expect, and accurate (or reliable) information is scarce. It is characterized by personal and corporate networks and relationships across importing, exporting, and transit countries. It includes global charities, and organized crime networks.

For example, emigrants – such as Nigerians in London, Ghanaians in Germany – collect or buy then ship discarded electronics back to those countries (Grant and Oteng-Ababio 2016). They are part of networks of small-scale entrepreneurs, along with those who ship cars, bikes, and other second-hand goods, but more vulnerable to prosecution (Lepawsky 2018). At the receiving end are brokers and companies who may (as in the case of China) be in fierce competition, driving prices up. In Ghana, the sector is also organized. There are several Ghanaian companies (registered in India or Saudi Arabia), and some foreign firms.

The e-waste recycling business has attracted global mining companies to sites in Africa. Belgium's Umicore, for example, has technology to refine and reprocess the ore after extraction. They ship the ore back to their refineries in Europe or North America for smelting and refining. In this way, Umicore has rebranded itself as a cleantech company, leaving behind a violent colonial past and a more recent past as polluter to become the world's largest recycler of precious metals. Other companies that now process e-waste include Glencore (Canada), Aurubis (Germany), Boliden (Sweden), and Japanese company Dowa (Knapp 2016, p. 3). Computer hardware companies like Dell and HP have established operations, and international NGOs are also entering the arena.

Governing Global Production, Disposal, and Flows of Discarded Electronics

E-waste – its production, collection, disposal, recycling, and trade – poses many governance challenges. The globalized e-waste economy magnifies these challenges. Different actors compete for potentially high profits, while the actual risks are transferred or distanced from point of use, often to vulnerable communities. Ironically, this means that e-waste risks are often displaced to where electronics were originally produced. Its production is dispersed, and recycling sites are highly mobile, therefore harder to control. Effective e-waste governance should get electronics producers to give up their very profitable business model, relying on planned obsolescence to drive ever-growing demand. But how?

By 2017, 67 countries – representing two-thirds of the world's population – had e-waste legislation. This number is up from 61 countries in 2014, but there are many gaps in both coverage and implementation, even in the wealthiest countries (Baldé et al. 2017). What we have today is a patchwork of official regulations, initiatives, and voluntary schemes, often limited by global national and local enforcement and implementation capacities. Efforts to build cohesive governance systems are hampered by the absence of a comprehensive global legal framework that harmonizes e-waste definitions and monitors or controls production and trade.

Effective governance of the production and fate of discarded electronics faces steep challenges. The e-waste stream is clogged with devices and appliances from an infinite number of sources. Its root causes – demand for technology, stocks of old appliances, planned obsolescence – are hard to address, even at the source. Asia is catching up with the US and the EU in total e-waste production (18.2 million tons in 2016), led by India and China. Africa was estimated to have produced 2.2 million tons in 2016, led by Egypt and South Africa. In Latin America, average per capita production in 2016 was 7 kg, low compared with the US and EU (11.6 and 16.6 kg respectively) but higher than Asia (4.2 kg). Production there is led by Brazil and Mexico, where cellphone penetration has been highest.

Extended Producer Responsibility
Dominant domestic e-waste governance approaches to date have placed responsibility and/or accountability on the producers and retailers of electronic goods. The concept of *extended producer responsibility* (EPR) has become a mainstay of electronics' governance. Swedish economist Thomas Lindhqvist defined EPR as a strategy to decrease the total environmental impact of a product, by making manufacturers responsible for the entire lifecycle of their products, especially for take-back, recycling, and final disposal. EPR has driven e-waste regulation in the EU, China, and California, and has been adopted in Kenya, Nigeria, and other developing countries.

The EU has made the most progress in developing cradle-to-grave electronics regulation. Two ground-breaking directives – the Waste Electrical and Electronic Equipment (WEEE) Directive and the Directive to Restrict the Use of Hazardous Substances in Electrical and Electronic Equipment (RoHS) entered into force in 2003. The WEEE directive sets strict targets for producer take-back of used electronics and RoHS restricts the use of hazardous substances in electronics manufacture. From 2019, EU member states are required to show collection of 85 percent of discarded electronics. Producers pay for collection and disposal, not households, and penalties imposed by member states are expected to be "effective, proportionate and dissuasive" as laid down in the Directive. StEP's Person in the Port project (see earlier in chapter) found that much of the e-waste smuggled in used cars to Nigeria was from Europe. The charity exception also demonstrates that EPR is quite porous. E-waste escapes.

China is the largest e-waste producer in the world, with 7.2 million tons in 2015 (Baldé et al. 2017). Its own EPR system, though much weaker than in the EU (Reagan 2015), makes steps toward a country-wide legislative framework. India enacted EPR legislation in 2011. Nigeria, Kenya, and Ghana introduced bills in 2016.

In the US the only federal law on the books regarding e-waste designates cathode ray tubes as hazardous waste due to leaded glass, but has few restrictions on e-waste exports, especially if labeled for recycling. It has not ratified the Basel Convention. Many of its states, however, make their own e-waste and electronics regulations. California's Electronic Waste Recycling Act (2003) allows retailers to collect fees at point of sale to take used electronics back for recycling. As of 2018, 25 states and the District of Columbia had some form of e-waste legislation on the books, according to the National Conference of State Legislatures. In the

2000s, it emerged that much e-waste recycling in the US was done by prison laborers, a strongly criticized practice (Sheppard 2006).

In principle, EPR should incentivize better product design, thereby diminishing excess waste production. Manufacturers bear take-back costs and penalties for violating their responsibilities. In practice in the US, Canada, and EU, designs have become less durable and the costs of complying with EPR responsibilities are passed directly to customers (Lepawsky 2018, p. 168).

EPR, as with other voluntary governance initiatives, works best when backed by strong regulation and monitoring capacities. Sometimes EPR programs are run by governments, but there are many examples of shared responsibility between government and third-party organizations that monitor or certify manufacturer compliance but have a harder time strengthening and enforcing initiatives. Producers need to be incentivized to comply, and less well when producers can move their basis of operation – production, sale, or disposal – to avoid regulations.

Effective EPR requires transparency, enabling, for example, NGOs to "name and shame" big-name producers, like Dell, Samsung, or Apple. In developing countries, where products often come with no brand name attached, this is much harder to do. Also, as with clothing and shoe production, the big brands outsource manufacturing to companies based in Asia, Latin America, or elsewhere. The Chinese company FoxConn makes electronics for Apple, but also for Amazon and other competing brand names, and there are other such companies. In such complex, rapidly shifting global supply chains, there is little space for product redesign and assigning accountability. Rapid innovation cycles make it hard for regulators to keep up, and supply chain complexity undermines transparency (Gardner et al. 2018).

Enforcement and Policing
A second component of e-waste regulation remains *enforcement*, including policing of criminal activity (smuggling, labor exploitation, materials theft), controlling borders (as much as possible), and capacity-building. The same groups who smuggle wildlife, drugs, and weapons are also likely to smuggle e-wastes (Elliott 2014; Gibbs et al. 2011). The UN/Interpol funded Green Customs Initiative works to train and fund inspections authorities in ports to catch and stop environmental crimes. Interpol's environmental crimes unit has carried out operations to stop waste smuggling: in August 2017 it announced it had seized over 1.5 million tons of illegal waste (including electronics and metal, most of it connected to the automotive industry) in a 30-day global operation (Boteler 2017).

Governing the Trade
Although the Basel Convention is the only binding international legal framework for controlling the international trade in electronic wastes, it has serious flaws. It can only regulate materials that are shipped as hazardous wastes, which most e-wastes are not. It has no provisions that allow it to regulate South–South or South–North trade. Its 1995 Ban Amendment only forbids export from these countries to non-OECD countries and non-parties to the Convention and has not entered into force. Therefore, the Basel Convention has not been able to incorporate effective governance of the e-waste into its mandate, although it has provided guidelines and partnerships with the industry. Other efforts to govern the trade also focus on bans. The EU's WEEE Directive restricts export beyond its borders. China banned imports in 2002 although with gaps in implementation. Estimates from EU and UN studies suggest that at least 8 million tons have been

smuggled in annually (Geeraerts et al. 2015; UNODC 2013).

The trade in used electronics has not been stopped – and perhaps should not be, if e-wastes are indeed to be recycled efficiently in significant quantities and used beyond their point of initial discard. This is not to underestimate labor conditions, environmental hazards, or criminal organization in the actual shipping of wastes. These governance challenges have generated innovative responses, the subject of the next section, towards building a new global green economy through enabling refurbishment and repair.

Repair, Reuse, or Disposal? Building the "Informal Green Economy"

In 2016, China-based journalist Adam Minter reported on visiting the Huaqiangbei District of Shenzhen in Southeast China, home to a vast used electronics shopping mall, where refurbished circuit boards, computers, phones, and so on are much in demand (Minter 2016). This mall reflects an industry that has become part of the mainstream marketplace of the world's largest country. It is potentially, as Minter describes, "one of the most powerful informal green economies in the world." The studies by Minter, Lepawsky and others have started to reveal the repair economy that is also a vital part of the global e-waste economy.

The reality of dealing with used electronics in developing countries has always been more complex than assumed. Coverage so far of the e-waste trade has underestimated how many of the used electronics shipped to Southern countries are repairable, and not simply dumped. Used appliances are bought by local people, including officials and small business owners. In Ghana, computers go to schools, and televisions to homes: "without the televisions they have repaired over the years, nobody would have built TV towers. And the same goes for internet access" (Robin Ingenthron, of Fair Trade Recycling, quoted in Ottoviani 2015). These practices fit with far higher rates of recycling reported in developing countries compared with developed ones (Minter 2013). While in the end devices can no longer be repaired, e-waste itself is unavoidable but under the right conditions, it can be put off and reduced.

Descriptions of Agbogbloshie describe dangerous conditions but not a burning wasteland, a "hell on earth." Rather, it is the end point for electronics that Ghana's small-business recycling sector has been unable to reuse or repair. In a 2015 investigative piece for Al Jazeera ("E-Waste Republic"), Jacopo Ottoviani maps a more complex layout of the area, with repair shops and markets located between the port and the dump yard, before the breaking down, extraction, and disposal of the remainder happens.

The Chinese government has focused on cleaning up its informal recycling centers. By December 2015, Guiyu had transformed, built into an industrial park, albeit still with a toxic legacy, including lasting toxins in the soil in and around the city. Thousands of workers were also displaced in this process, as reported in the *South China Morning Post* in September 2017 ("China's most notorious e-waste dumping site now cleaner, poorer").

If experience to date with e-waste and the e-waste trade (and the other cases in this book) is anything to go by, the most effective governance approaches are those that take the inherent value, as well as the risks, of these discards seriously. Alternative policies have proliferated in recent years. One, the Best of Two Worlds (B02W) has been put forward by the StEP Initiative (2010). The StEP (Solving the

E-Waste Problem) Initiative is a multi-stakeholder initiative run by the United Nations University, including international agencies, companies (such as Dell, HP, Microsoft, Umicore), government agencies, and universities. This proposal advocates taking advantage of low-cost labor in developing countries to dismantle electronics, and high-quality technology in developed countries to reprocess the extracted ores with minimal environmental cost.

The problem with the B02W approach is that it runs the risk of reproducing what primary resource trade policies have done through the ages: the displacement of value to the wealthy jurisdictions. As with timber from Southeast Asia or tuna from the South Pacific, communities at the site of extraction lose access to the value added along the e-waste supply chain. How can we stop this from happening and allow communities in developing countries to benefit?

Lepawsky et al. (2017) suggest, as a direct alternative, Ethical Electronics Repair, Reuse, Repurposing and Recycling (EER4), a program that opens the door to a certified fair trade in e-waste that includes fair wages and working conditions. Ethical trade is a radical idea next to most of the work on the e-waste trade but is also being developed by industry associations with respect to plastic scrap (see chapter 6). It is a necessary component to governing the e-waste trade but must apply lessons from other types of certification schemes.

Certifying that commodities meet predetermined standards of sustainability, labor rights, and so on, is not a new idea. Timber, coffee, cocoa, fish all have associated certification initiatives that are well-respected and under certain conditions work well, but have overall underperformed (Cashore et al. 2004). Usually these standards are applied at one end of the transaction, to exporters and producers.

If these actors are in developing countries, it is harder to verify and monitor these standards. The e-waste case is different. Exporters of used electronics should certify to the safety or usability of the devices they ship. Importers then must certify that electronics are being recycled or dismantled under safe conditions. The latter will be much harder than the former to verify. Further, these initiatives are also nearly all voluntary, which means they do not cover the worst actors in the system who choose to stay out. In the complex e-waste networks sketched above, this is likely to be the case as well.

Maintaining value in electronic wastes while reducing risks is one important platform of effective governance. So is product design. Indeed, the two are linked. Better product design, that enables a longer life, repair, and lower risks while dismantling end-of-life devices contributes to a more equitable global waste and recycling economy. However, it is also treading on the toes of manufacturers and some in the recycling industry, who do not want the disruption of business as usual. However, repair and recycling businesses North and South are under threat as our ultra-thin, non-repairable gadgets make their way through stages of production and use with ever shortening lifecycles. These manufacturing practices – planned obsolescence on steroids – not only rob recyclers of jobs, they rob people in developing countries of the opportunity to buy affordable refurbished electronics.

Contrary to expectations of EPR programs, companies have designed electronics to be harder to fix than previous generations, and manufacturers no longer release repair information or authorized parts (Matchar 2016). Copyright laws make unauthorized repair of and unlocking devices an infringement of corporate intellectual property. A growing "Right to Repair" movement in the US, Europe, and elsewhere has started to contest these corporate practices. Farmers are credited with taking an early lead: as farm equipment becomes more computerized, it is harder to fix when it breaks down, immediately costing agricultural production.

The Right to Repair movement encompasses diverse tactics and initiatives. One is to push for legislation to require producers to make it possible to repair or unlock electronics. "Fair Repair" Bills and ballot measures in US states are spearheaded by the Repair Association. In France, planned obsolescence is punishable by fines. Another part of the movement is about setting up small stores and enterprises to repair or teach people to repair anything from smartphones to shoes. The Swedish government gives tax breaks on repairs. From Stockholm to Brooklyn, this is an act of resistance to electronics manufacturers and the broader culture of disposability.

In the end, the key is understanding what sorts of wastes are being imposed on people who depend on recycling for their livelihoods, and what sorts of wastes are creating high risks and damages that need to be prevented or "designed out" in the first place. For example, imagine electronics being designed for easy disassembly, having their toxics sharply reduced, and still providing a (fairer and healthier) livelihood for informal recyclers. As a movement, Right to Repair has touched a chord with consumers across political divides and is utilizing diverse tactics and targets to meet its goals (Economist 2017; Mitchell 2018). However, industry has mobilized to oppose measures regarding electronics manufacture, which is also part of a broader struggle over the ownership of digital resources (the songs on your iPod, or books on your tablet). These conflicts are set to become more highly politicized over the next few years.

Conclusions

Discarded electronics are a cornerstone of the global waste economy. They exemplify a global resource frontier and the magnified risks that characterize it. Changes in how this economy works, especially through the trade, have upended conventional assumptions about the North as perpetrator and the South as victim. We see this most clearly by focusing on the demand for used electronics and the complex trade routes they take. Even understanding, however, that more electronics are repaired or refurbished than previously thought does not negate the serious risks faced by workers in dismantling used electronics, hence their framing as electronic wastes remains valid. Extracting urban ore is a multinational business also open to worker exploitation.

Governance challenges along the electronics supply chain are myriad and multifaceted. This chapter focused on the secondary resource extraction, trade, and final disposal ends of the supply chain. By the end, however, product design, at the start of the supply chain, takes on critical importance, even for global waste governance. The "right to repair" in the face of proprietary hardware is a rallying cry for organizations around the world.

The next chapters continue these themes, although in different ways. Food waste (chapter 5) is less of a globally networked commodity but displays levels of activism and governance entrepreneurship that have made it more tractable than other global waste problems. Plastic scrap (chapter 6) brings us back squarely to a central problem with the e-waste trade: how to distinguish "waste" from "scrap" at the global level. It continues the global governance discussion, focusing on specific global venues where these decisions could be made.

CHAPTER FIVE

Food Waste

In May 2017, an article in *The Guardian* newspaper called leafy greens the "tip of the food waste iceberg" (Carlin 2017). The UK supermarket chain Tesco revealed in a 2017 study that consumers throw out 40 percent of the bagged lettuce they buy each year, the equivalent of 178 million bags (Smithers 2017). This problem is not confined to consumers. Supermarkets themselves throw out yet more bagged lettuce even before it reaches the sell-by date printed on the package. In the US, supermarkets discard 10 percent of their stocks, according to a 2012 study from the Natural Resources Defense Council (NRDC; Gunders 2012, updated in Gunders et al. 2017). Whole harvests can be destroyed before reaching supermarkets following outbreaks of food-borne diseases such as *E. coli*. Leafy greens exemplify problems across food supply chains that contribute to global food loss and waste, estimated by the United Nations Food and Agriculture Organization (UNFAO) to add up to one-third of the world's total food production.[1]

Food waste and loss became high-profile political issues across the industrialized world in the late 2000s. They have a global footprint. Food waste generation and its impacts are visible from the household up to the largest agricultural areas. It connects with individuals' concerns about consumption and waste practices and is shaped by global trade and aid policies. Tracking food across its supply chain connects waste with broader narratives of food production and politics and with contemporary concerns about food security, safety, and sovereignty.

The landscape of food waste-related organizations, campaigns, and coverage is overwhelming. Celebrity chefs, mainstream NGOs, and UN agencies are just a few of the types of actors involved. So are local charities, food banks, and urban farmers. In 2015, the UN started to implement globally agreed upon Sustainable Development Goals (SDGs). Goal 12 (Responsible Production and Consumption) aims to "halve per capita food waste at the retail and consumer level and reduce food losses along production and supply chains by 2030."

Food waste and loss have squandered reusable resources along entire global product supply chains. The extent and global ramifications of food waste in developed countries has magnified global risks. It affects food security in developed and developing countries and exacerbates global greenhouse gas emissions. Yet, it is also a problem with feasible solutions. Unlike plastic scrap (chapter 6), it is quite readily returned to productive uses (composting, donation, and redistribution) and there are many options for preventing it in the first place. There is, as near as possible, universal support for tackling the problem.

Food waste, broadly defined, illustrates waste as *inputs* (see box 1.1). Zero Waste and circular economy programs almost all include food waste. Whether these plans include biomass projects for generating energy or social enterprises to distribute non-perishable discarded food to the poor and homeless, food waste can usually be handled locally and safely, unlike either e-waste or plastic scrap. However, questions remain around policy choice, underlying values, and effectiveness. For example, composting is an effective and easily deployed way to dispose of discarded food, but not as good overall as preventing food waste in the first place.

This chapter opens by defining food waste and loss. It then analyzes how food waste gained prominence as a twenty-first-century problem. We examine its history, global variance, and causes as a food problem and as a waste disposal problem. Addressing food waste has generated governance innovations and initiatives, from private and public sectors. Food waste activism spans the largest NGOs to the urban poor, a movement that has been reproduced in efforts to reduce plastic waste production. We look at examples of food waste activism and governance efforts, from the individual to the global, from social supermarkets to date labels.

Finally, this chapter turns to the global political economy of food waste. First, contrary to mainstream assumptions, studies show that food waste is prevalent even in poor communities in the global South. Second, waste from foodstuffs exported from South to North shapes food sovereignty in the exporting country in subtle but significant ways. Third, food waste shipped as agricultural surplus in the opposite direction, from North to South, has shaped food sovereignty and security in less developed countries. This problem has been addressed in work on food, agriculture, and development aid, but adds insight into the impacts of food waste in developed and developing countries alike.

Food Waste and Food Loss

Waste occurs in food production, handling and storage, processing, distribution and sale, and consumption. Harvests are lost to weather or to price collapses. Consumers and retailers discard produce when it spoils, it does not meet taste or appearance standards, or there is too much of it. If it fails to meet official health and safety standards, food is destroyed. Much discarded food from households and restaurants winds up in landfills, excluded from productive reuse.

The UNFAO, on its food loss and waste web portal at fao.org, distinguishes between food loss and food waste:

> Food that gets spilled or spoilt before it reaches its final product or retail stage is called *food loss* ... Harvested bananas that fall off a truck, for instance, are considered food loss. Food that is fit for human consumption but is not consumed because it is or left to spoil or discarded by retailers or consumers is called *food waste*.

The Waste and Resource Action Programme (WRAP) defines food waste as "any food (or drink) produced for human consumption that has, or has had, the reasonable potential to be eaten, together with any associated unavoidable parts, which are removed from the food supply chain."[2]

Yet these definitions easily falter, especially when considering the contextual dimensions of food waste. "Edible" is culturally dependent. What might be considered unfit for human consumption by some – e.g. whale blubber or fish heads – is food for others. So, fish heads thrown away in a London market might be perceived as a waste of a precious commodity by a local immigrant (Soma 2017, p. 1449). Likewise, "removed from the food supply" may or may not mean being returned to the soil, as when crops are ploughed under and provide nutrients for the next planting. This may not be the optimal use of these crops but is it still waste or loss?

Box 5.1 outlines some of the key statistics on food waste provided by the UNFAO and other organizations. In a review piece, Li Xue and colleagues go through existing

Box 5.1: Statistics on food loss and waste

- Roughly a third of the food produced in the world for human consumption every year – approximately 1.3 billion tonnes – is lost or wasted.

- Food losses and waste amount to roughly US$680 billion in industrialized countries and US$310 billion in developing countries.

- Fruits and vegetables have the highest wastage rates of any food: The National Resources Defense Council (NRDC) estimates that 52 percent of fruit and vegetables produced or purchased in the US, Canada, Australia, and New Zealand is discarded.

- Global quantitative food losses and waste per year are roughly 20 percent for meat and dairy, and 35 percent for fish.

- Per capita waste by consumers is between 95 and 115 kg a year in Europe and North America, while consumers in sub-Saharan Africa, South and Southeast Asia, each throw away only 6–11 kg a year.

- In developing countries 40 percent of losses occur at post-harvest and processing levels while in industrialized countries more than 40 percent of losses happen at retail and consumer levels.

- The carbon footprint of wasted food is estimated at 3.3 gigatons of carbon dioxide equivalent.

- The production of wasted food uses around 1.4 billion hectares of land, or 28% of the world's agricultural area.

- It also wastes around 250 km^3 of surface or groundwater ("blue water"), more than 38 times the blue-water footprint of US households.

Source: UNFAO's Save Food Global Initiative on Food Loss and Waste Prevention; also based on UNFAO (2011, 2013), Gunders (2012) and Gunders et al. (2017).

studies and data on food loss and waste, finding that many are based on secondary sources (Xue et al. 2017). They argue that a full audit of food waste data is needed in order to craft good policy solutions.

Food Waste on the Political Agenda

The issue of food waste rose up political agendas in the global North in the late 2000s and 2010s, rapidly gaining visibility and salience. We care about food waste because of its impacts on food security and climate change, and its connection to the politics of food and broader social norms. It is an issue of global political economy because of how it extends across food supply chains and continents. We also care about it because it hits us where we live, touching what we buy and eat and throw out every day, unsettling the long-held social norm of thrift.

The contemporary era of economic globalization, commodification, and consumption began in the early 1990s. Consumer wastes – food, plastics, electronics, clothing – began to pile up alongside other types of municipal solid waste. Such visible wastes connected to debates over individual responsibility and global sustainability. Food waste entered already raging food politics debates in the global North in the 2000s (Clapp 2016; Nestle 2002). The use of pesticides and genetically modified organisms (GMOs) in food production helped fuel local food and organics movements. So did the concentration of the agricultural sector in the hands of a few multinational corporations. Contemporary concern for food waste in developed countries also coincided with a wave of concern over obesity. Urban food deserts and food insecurity were also on activist and political agendas.

These developments provided fertile ground to broaden food politics to include food waste. The global financial crisis of 2008 was a wake-up call (Evans et al. 2013). Rising food prices, caused in part by a sudden demand for biofuels, brought home to many the vulnerability of the food supply. Biofuels, made from crops including grains and sugarcane, can substitute for fossil fuels, but to grow them in adequate quantity means displacing food crops. *Food security*, according to the UNFAO and others, exists when all people, at all times, have physical and economic access to enough safe and nutritious food that meets their dietary needs and food preferences for an active and healthy life. Redirecting food waste to food banks or "social supermarkets" is seen as a way of increasing food security, at least in the short term.

Food waste in different forms, as the final section of the chapter discusses, can also affect food sovereignty. *Food sovereignty* is "the right of peoples to healthy and culturally appropriate food produced through ecologically sound and sustainable methods, and their right to define their own food and agriculture systems" (Declaration of Nyéléni, the first global forum on food sovereignty, Mali, 2007). Exports of produce and other foodstuffs from southern countries to northern markets is an unlikely example (see below).

Food waste also contributes directly to climate change through methane emissions from rotting biowaste in landfills. Indirectly, it contributes through wasted transport, production, and processing-related emissions. According to the UNFAO's 2013 report, if food waste were a country it would be "the third largest greenhouse gas emitter in the world" after the US and China. Combatting food waste is seen increasingly in global climate politics as a way to cut greenhouse gas emissions (O'Neill 2019).

These trends provided the impetus for activists and policy entrepreneurs to raise concerns, devise solutions, and push for political action over food waste, especially in the global North. The next sections examine the history and causes of food waste, then the emergence and role of food waste activism.

A Lengthy History

Food wastage and loss have deep historical roots.[3] Social histories of food waste relate how in the eighteenth and nineteenth centuries European nobility saved banquet leftovers for the poor, who they believed to be more immune to the effects of spoiled food (Schneider 2013). As urbanization intensified, and people lost easy access to fresh food, food spoilage became more of a problem. Canning technology was first developed in the early nineteenth century in response to these demands, but not without early disaster (Geoghegan, 2013). It was almost assigned to the trash can of history in the middle of that century due to malfunctions, rotting contents, and subsequent deaths (especially on long sea voyages). It survived and went on to become the first form of large-scale food preservation, with mechanized production starting in the late 1800s. Canned food supplied armies in World War One and beyond, and condensed milk and baked beans became household staples. Now households in Europe and the US collectively consume 40 billion cans of food annually, even with alternative storage technologies.

Food loss and waste on a large scale emerged with systems of industrialized agriculture and long-distance transportation of foodstuffs in the years following World War Two. Government policies – especially in the US and Europe – led to mass-produced and cheaper food for their populations, and the start of massive surpluses of grain and other foodstuffs.

Europe's Common Agricultural Policy (CAP) hit the headlines in the 1980s as an early example of food waste on a large scale. First implemented in 1962, the CAP was designed to protect farmers in the European Economic Community (now the European Union) and to maintain lower prices within the trading bloc. It had, however, the unintended effect of creating food surpluses. In the 1980s, it became a figure of derision for Euro-skeptics for its "wine lakes" and "butter mountains" (sadly not actual lakes and mountains), symbolic of the waste generated by a bureaucratic system insulated from the "efficiency" of the free market. Wine was destroyed or turned into vinegar. Butter and grains were destroyed, used as food aid, or dumped on developing country markets at low prices. The EU did, however, later lead the way in the fight against food waste: its 1999 Landfill Directive required its member countries to reduce biodegradable municipal waste sent to landfills to 35 percent of 1995 amounts by 2016, including food waste.

The influence of the norm of thrift has waxed and waned over these years. According to the *Oxford English Dictionary*, thrift is the quality of using money and other resources carefully and not wastefully. It is a powerful norm around food use that has endured for hundreds of years, driving the organization of households in good and bad times. Cookbooks and domestic science courses reinforced thrifty practices in the nineteenth century (Evans et al. 2013). The impacts of the Great Depression created a generation brought up under scarcity. In 1917, Herbert Hoover, in charge of the new US Food Administration founded the "Clean Plate Club." Food scarcity and rationing during World War Two turned food waste into a national security issue in the Allied nations ("Food is a Weapon – Don't Waste It").

Since then, the norm of thrift has weakened but not died. Food is abundant and cheap, and we are far removed from its production and disposal, but uneasiness remains: "most citizens inherently perceive that food, the very essence of life, is handled wrong" (Aschemann-Witzel et al. 2016, p. 281). Campaigners use this uneasiness in their battle to stop consumers and retailers wasting food. Still, calls on individuals to embrace this "Protestant ethic" may fall flat. Laying blame or imposing guilt on consumers, especially when structural factors outside consumers' control shape their actions, can be futile (Gjerris and Gaiani 2013).

Causes and Regional Variations

Food loss and food waste look different across the world's major geographic regions. Food waste through consumption dominates in Europe, North America, and East and Southeast Asia. Food loss at production and distribution stages is most significant in sub-Saharan Africa, Latin America, and South Asia.

The causes of food waste and loss in developed countries have been studied extensively (Evans et al. 2013; Gunders et al. 2012, 2017). For consumers and households in the US and other wealthy countries, food is cheap, especially relative to average incomes. Restaurants serve huge portions, diners take leftovers home and store them in fridges already stuffed with groceries. For retailers, sell-by dates on food (see box 5.2), regardless of whether they reflect actual spoilage, mean much is thrown out before it needs to be. The position of the agri-food industry and retail sector can be two-sided. While on the one hand visibly embracing actions against food waste, it and some retailers are concerned that campaigns and legislation are reducing their profits. Bulk discounts – which have been shown to increase food waste – continue as standard practice.[4]

Cosmetic imperfections lead to discard of tons of otherwise edible produce. One study found that almost 90 percent of tomatoes at a commercial farm in Queensland, Australia, were thrown away based on appearance, a number thought to be high but typical (McKenzie et al. 2017). Crops affected by pests, drought, or storms rot in the fields if there is no one to harvest them. Farm labor is a problem; labor shortages in the agricultural sector is another reason crops go unharvested. Immigration crackdowns in the US have led to worker shortages that cut harvests and leave produce in the fields.

A larger narrative connects back to the leafy greens story that opened this chapter: many perceive a tradeoff between sustainability (less wastage) and safety (including public health). Worries about the health and legal liability implications make producers and retailers discard food more readily. The plastics industry touts its role in preventing food waste through packaging. Even though consumers want to avoid throwing edible food away, in the moment they prioritize taste, convenience, or health concerns (Aschemann-Witzel 2016, p. 409).

Studies of developing countries focus on food loss at the production end of the food supply chain, which far outweighs food waste downstream by retailers and households. For example, sub-Saharan Africa and South Asia may each lose about one-third of food produced in handling and storage, and proportionately more in processing than other regions. According to the UNFAO, food loss in developing countries can be traced back to financial, managerial, and technical constraints in harvesting techniques as well as storage and cooling facilities. Many Southern countries do not have continuous cooling infrastructures that cover processing facilities, transportation, and the wholesale markets where most produce is sold. Rice is a good example. Nigeria and Bangladesh are two of the largest global producers of rice – but post-harvest losses (25 and 12 percent respectively) lead to food insecurity and import dependence (Kumar and Kalita 2017, p. 6). Below, we look at studies that question this prevailing view, and introduce an alternative global perspective on food waste.

Food Waste Disposal

Waste disposal facilities and policies have failed to keep up with the scale and scope of the food waste problem. This

is despite the relative ease and effectiveness of composting at household, municipal, or industrial scales, compared
with the disposal of other types of wastes. According to the
US EPA, food waste made up 20 percent of all trash that
went into landfill in 2012, the single largest component
of municipal solid waste streams. Further, "of all the food
that is lost at different stages from farm to fork, only 3 percent is composted" and only about 10 percent is recovered
(Gunders 2012, p. 14). This is despite the fact that there
is no shortage of composting technologies, from pigs to
anaerobic digesters. However, setting up equipment, collection, and transportation infrastructures at curbside and
factory levels is expensive.

At the community level, individuals and families face
challenges in composting food waste. For example, few
municipalities in the US provide curbside compost pick-up
that includes food waste, and only a handful mandate this
practice. Education on what, and how, to compost wastes
is lacking, as is attention to where and how people live,
and contamination rates continually threaten the viability
of compost collection. Peeking into compost bins on my
campus always reveals a whole range of non-compostable
items (including plastic). The contents of these bins are
thrown into the regular trash. Apartment dwellers with no
access to the outdoors find composting particularly challenging, and landlords may not supply a separate green
waste bin. In the EU, there are a greater number of mandatory composting programs for food and other biodegradable
wastes (see below). In the retail sector, wasteful disposal
practices include using dumpsters rather than composting facilities for discarded foods. Some food retailers toss
bleach onto discarded but edible food to deter dumpster
divers and destroying food under recall.

Food disposal practices and safety regulations make

recovery and re-sale or donation of discarded (or out-of-date) items hard, exacerbating wastage. Traditional governance of food – local, national, global – has been driven by food safety and public health, to avoid disease outbreaks and deaths due to large-scale food poisoning or spoilage. Given the world's experience with major food scandals, this is a reasonable choice. In Scotland, animal carcasses are treated as hazardous waste following the Mad Cow Disease crisis in the UK that started in the 1980s. There are laws against donation of discarded (but still edible) food, especially perishable food to charities in Europe and other parts of the world. There are very few "Good Samaritan" laws that explicitly address food donations. The 1996 Emerson Food Donation Act in the US is the only prominent example of such a law: it protects good faith food donors from liability in the case of harm to recipients of donated food.

#nofoodwaste:
Activism and Policy Entrepreneurs

Food waste activists have played a critical role in pushing the issue onto political agendas and raising concern and awareness among the broader public.[5] Celebrity chefs like Jamie Oliver have lent their voices to the cause. Of the cases in this book, food waste has generated the most diverse activist groups and organizations, all along the supply chain, and some of the most inventive campaigns. These have included art installations and pop-up dinners made from rescued food. Lessons from this movement's success have informed the strategies and repertoires of oceans plastics activists (see chapter 6).

Local community organizations, student groups, farmers and other groups encourage food waste prevention, and redistribute discarded foodstuffs or cosmetically flawed produce. They work with authorities to create or change local laws and ordinances around food waste, and partner with local businesses. One of the oldest, New York City's City Harvest, was founded in 1982. Other food waste NGOs focus their activity on research, raising concerns and actively lobbying political and business actors for policy change. WRAP set up its Love Food, Hate Waste Campaign in 2007. Save Food is a global initiative in partnership with UNEP, UNFAO, and others. ReFED is a California-based think-tank that works with businesses, non-profits, and others to develop and assess potential solutions to the food waste problem. It also maintains a map of food waste "innovators" – activists, technology developers, entrepreneurs, and others.[6]

Large international NGOs, such as the National Resources Defense Council, the World Resources Institute, and Oxfam have added food waste campaigns to their portfolio, as have major food-related NGOs such as Food First. Both the NRDC and the World Resources Institute have produced influential work on food waste.[7]

The Freegan movement embraces dumpster-diving, drawing attention to the quantities of edible food discarded by food retailers every day (Barnard 2011; Wittmer and Parizeau 2016). It provides a visceral example for observers to understand urban food security and waste. Dumpster-diving for food thrown out by restaurants and supermarkets has become an act of resistance, especially in the face of locked dumpsters and chemically contaminated food. Freeganism's historical antecedents include the San Francisco Diggers movement of the late 1960s, for whom giving food away and establishing free stores was part of their anarcho-syndicalist performance repertoire. It echoes food sovereignty organizations in its call for broader political reforms.

Food waste activism has driven policy breakthroughs, which in turn have mobilized more activism. The EU Landfill Directive of 1999 provided part of the impetus for the founding of WRAP in 2000 as an NGO with government funding. The USDA's 2015 initiative to halve food waste across the US by 2030 mobilized stakeholders across the supply chain to reduce the amount of food entering the waste stream (Musulin 2016). As a social movement, food waste activism has changed policy, created alliances, and mobilized support across traditional fault lines. Food waste is not the hardest issue to gain support for. People agree on the scope of the problem and that something must be done. It is still a complex, diffuse problem that is hard to rein in (Morath 2018). Activists have combined research, awareness-raising and advocacy, built allies in government, retail sector, and food and waste industries, and utilized social media to get their message across.

Food Waste Governance

Addressing food waste has generated a profusion of governance experiments and innovations across scales. As with the problem itself, these initiatives exist across jurisdictional levels, and at different points all along the supply chain. They focus on both the prevention and reuse of food waste: composting, food donation, education, expiration dates, and packaging. Large-scale bio-waste disposal includes energy generation through methane capture or anaerobic digestion.

The food supply chain in its conventional form exemplifies the linear economy, from farming practices that strip nutrients from the soil to the large-scale waste of edible foodstuffs. Food waste policies and initiatives are an effective way to implement circular economy policies and practices. Practices at the end of the food supply chain may then feed back into the complex industrial agricultural systems that dominate food production. Policies designed to divert food waste from landfill are an important plank of many municipal Zero Waste policies. London's Zero Waste plan targets biowaste first and foremost.

These initiatives do not need a global or even national reach to have an impact on the bigger problem, and local initiatives can add up to address food waste on a larger scale. Household composting and local and centralized composting facilities divert food waste back into a usable resource. Local charities and other groups can capture edible food and resell or donate it (sometimes). Food loss is harder to address locally.

Community-Based Initiatives
Local organizations and community groups have partnered with restaurants and supermarkets to collect still-edible leftover food or close-to-expiration groceries and produce, to redistribute it to the homeless and to food banks. They have developed consumer education campaigns. College and institutional cafeterias have adopted tray-less food service, composting bins, and posters and displays.

"Social supermarkets" combat food waste and food insecurity by selling slightly damaged or close to out-of-date groceries at low prices. The first of these were opened in France in the late 1980s, and as of 2016, over 1,000 existed across Europe. "Ugly veg" or "imperfect produce" (also known as "wonky," "inglorious," "gorgeous on the inside") campaigns have gathered steam (Bhatia 2016). Stores, farmers' markets and online enterprises specialize in selling and delivering imperfect produce.

City and State-Level Initiatives
Municipal and state-level initiatives around food waste concern production, collection, and composting of biowastes (food waste and green waste, such as garden wastes). San Francisco, New York, Vancouver, Portland, and Seattle have household composting mandates, and over 150 communities in the US, mostly small cities and towns, have composting schemes (Levitan 2013). San Antonio, Texas, is an example of one larger city with collection. Washington, DC, also announced plans in 2017 for its own composting infrastructure.

In the EU, 14 member states have door-to-door bio-waste collection (food plus garden waste) as do 19 of their capital cities. Paris established such a system in 2017. Another five members have civic amenity sites for bio-waste, although in a few cases, collections are monthly rather than weekly (BIPRO/CRI 2015).

National and Global Initiatives and Regulatory Frameworks
Taking the lead, France passed a law in early 2016 forbidding supermarkets to throw away food, instead donating it to food banks. The law also lowered legal barriers for stores to give their food away to charity. Other countries, states, and cities are considering similar laws or voluntary programs. In 2014, Singapore's National Environment Agency launched an educational campaign ("Love Your Food") targeted at consumers, schoolchildren, and retailers. The EU's circular economy strategy makes food waste and loss a priority.

Many of these initiatives are propelled by the UN Sustainable Development Goals (SDGs). The 17 SDGs adopted by the UN and its member countries in 2015 address global poverty, education, gender equality, water, food, climate change, among others. They reflect a global target of achieving global sustainability by 2030. They are not legally binding. However, they provide a normative framework and ideas for local and national actors to set and meet their own goals (www.un.org/sustainable development). The US Department of Agriculture and the Australian government have each adopted these targets.

The US waste management industry plans to increase its composting and food recycling capacity. The Blumenthal–Pingree Food Recovery Act (introduced in 2015) aimed to reduce food waste across the board. It provided a comprehensive framework to tackle food waste. Measures include incentives for schools to buy imperfect produce for student meals and funds to construct large-scale composting and anaerobic digestion facilities for energy production. It also contained measures to standardize date labels on food packaging (see box 5.2). Although it and similar measures failed to be enacted into law, they indicate growing support for doing something to reduce food waste (Gunders et al. 2017, p. 6).

Fewer initiatives exist in developing countries, especially the poorest countries. Where food waste initiatives do exist, they are targeted at the growing middle classes (for example in India and South Africa), although ReFED has also mapped food bank initiatives in Malaysia and Nigeria. The UNFAO's Initiative on Food Loss and Waste Reduction is a partnership with Messe Düsseldorf (one of the world's largest trade fair corporations). One of the biggest needs is cooling technology for food transport. The push for such technologies has, however, come from below, from farmers, not from the companies who grow food or import it from countries such as India (Earley 2014).

The technologies that can make the most difference in food waste diversion are centralized composting and anaerobic digestion facilities. Anaerobic digestion

Box 5.2: "Use by," "best before," or "best if used by"?

Whether the label says "use by," "best before," "freeze by," "enjoy by," or something else, date labels on food are among the most frequent reasons why consumers and retailers discard uneaten food. These labels are put on food packaging by manufacturers. They create confusion because they are not standardized, and information about what they mean is hard to find. Do they refer to peak freshness or taste? Best appearance? Or do they mean the food is unsafe to eat? WRAP estimates that up to 20 percent of household food waste in the UK is due to confusion over date labels. A 2018 study from the EU estimates that 10 percent of EU-wide food waste relates to inadequate date labeling. Food producers and supermarkets use labels as a guide for re-stocking, often pulling food two or three days before the date. Many believe date labels are mandated by regulatory authorities, but with a few exceptions (such as baby formula), they are not. They are voluntary and meant as an indicator for flavor as much as spoilage. "Sell-by" is a label that matters only to retailers, and should not even be seen by the public.

Standardizing date labels could generate benefits of $1.82 billion annually in the US alone, as reported in *Food Dive* in 2016 (Heneghan 2016). Alternative date labels such as "best if used by," "use within x days of opening," "refrigerate after opening," "freeze by," labels that change color as food ages, or labels that identify products that are more dangerous if they spoil, may all increase clarity and change behavior.

The lack of standardization across jurisdictions suggests the need for legislation or at the very least, legislative guidelines. The Blumenthal–Pingree Act is an attempt to introduce federal guidelines in the US. The European Commission launched a sub-group of the EU on Food Losses and Food Waste in 2018. However, the slow pace of legislative action suggests changes will need to come from the private sector. Walmart's adoption of a single "best if used by" for its own-brand non-perishable foods could become the industry standard in the US.

Sources: NRDC/Harvard Food Law and Policy Clinic (2013), Gunders et al. (2017), and https://ec.europa.eu/food/safety/food_waste/eu_actions/date_marking_en

processes break biodegradable waste down into biogas and solids – the latter can be used as fertilizer, and the former can be used for heat or energy. While there are, for example, 2,000 anaerobic digestion facilities in the US, only 40–50 take food scraps, according to ReFED. Centralized composting facilities are expensive but exhibit economies of scale: the larger they are, the lower the cost per ton to treat food or biological waste. These numbers suggest the need for government investment in this infrastructure.

Non-Governmental Initiatives

Non-governmental initiatives are created and implemented by private actors, such as supermarkets, food producers, and NGOs. Retail chains like IKEA and UK supermarket chains such as Tesco's have set their own standards for reducing food waste. This is significant in the UK as supermarket chains there wield significant market power over food producers (Khalamayzer 2017). Also in the UK, the Courtauld Commitment is a voluntary agreement within the grocery industry to improve efficiency and reduce waste from their stores. It is funded by the UK government and administered by WRAP.

The Effectiveness of Food Waste Governance

How effective are all these measures? WRAP reported in 2017 that household food waste in the UK had risen between 2012 and 2015, despite high profile efforts, although this rise can be ascribed to exogenous factors such as population growth, and an economic rebound after the Great Recession. The scope of the problem means that communication and coordination is necessary all along the supply chain, from field to supermarket to household to centralized composting facility.

Deciding the effectiveness of any policy measure is not easy, as there are many ways to measure it. ReFED keeps an interactive visualization of 27 food waste initiatives, accessible on its home page, that compares them along these and other measures, including greenhouse gas emissions reductions and water savings. ReFED's data show that the same measures that are most cost-effective (such as consumer education) do not lead to direct waste diversion, while centralized composting and biogas facilities do. Similarly, such facilities create more jobs. However, the cost of building adequate composting infrastructure – primarily large-scale composting facilities – is high. To compost effectively (especially when compostable plastic utensils and similar are part of the stream) takes over six weeks. This is expensive for local authorities and requires more capacity to keep up with volumes of waste.

There are ongoing questions about the individual politics of food waste. As with other consumer wastes, moralizing turns away potential supporters and ignores the preferences and priorities individuals have in their lives. Food historian Rachel Laudan wrote on this topic in a March 2017 blog post, "Why I Happily 'Waste' Food."[8] Internalizing individual guilt is an exhausting way to live and work, as many environmental activists have also found out.

Tackling food loss is much harder. There are many ways to tackle food waste but addressing food loss presents a different set of challenges. Food loss on agricultural land, in transit and during processing, is also a function of complex agricultural systems. Decreasing food loss means fostering nutrient cycling, shortening transportation distances, and utilizing byproducts. These reforms need deeper levels of intervention in systems and practices of industrialized agriculture (Jeffries 2018).

The View from the South

The experience and actions of Southern countries have played little role in efforts to address food waste, which have been most prevalent in the global North. High-level conferences designed to bring together policy making and NGO elites to address this shared effort have paid scant attention to what we can learn from developing countries about food waste. The global reach of food waste and food waste politics is neither fully understood nor acted upon. This section touches on two themes that shape the global political economy of food waste and loss, and which have shaped food security and food sovereignty in developing countries.

First, food waste is not merely a byproduct of wealthy, developed world consumerism. Studies are finding that food waste – as conventionally understood – happens in poorer Southern communities too, often in cities, filling a gap in understanding the causes and impacts of food waste across the board (Soma and Lee 2016; Porpino et al. 2015). Second, food is a mobile resource: global trade in agricultural produce creates a massive and complex network that stretches around the earth and back (Clapp 2016). Although we rarely think about it this way, an international "trade" in food waste flows from North to South has diminished food security in the recipient countries. This story is well known in foreign aid circles, but far less so in waste circles – those who work with food waste or on the waste trade.

Household and Retail Food Waste in the Global South
Although studies remain rare, work in Brazil and Indonesia (and elsewhere) show that food wastage occurs in upper, lower-middle-class and working-class households across cities and towns. Tammara Soma, studying Indonesian households, shows how practices of "gifting" and "ridding" food generate complex patterns of shifting excess and risk to others (Soma 2017). The cases of "ridding" food to household servants is not necessarily something that would show up in Western studies (although it does connect to studies of the early history of food waste, see above). "Gifting" food and the display of such gifts by the recipient among the upper classes leads to more food than can be used by the

recipients. In Brazil, food wastage in open air markets and at supermarkets is an issue, along with the treatment of surplus food in low- to middle-income families (Henz and Porpino 2017). Again, for households it matters to have (and display or share) bounty and abundance matters, a theme common across cultures. Brazil is adopting many of the measures seen in the US and Europe, helped by burgeoning civil society networks and organizations.

Food Exports, Aid, and Waste
Discarded foodstuffs have shaped food security and sovereignty in developing countries as developed countries have dumped surplus harvest on markets in Africa and other parts of the world as food aid (Gille 2012). This includes the surpluses generated by the European Common Agricultural Policy, referenced above. This problem arose from the practice of tying food aid to domestic grain and other foodstuff production by the donor countries, as Jennifer Clapp shows in an insightful analysis of food aid (Clapp 2012). Starting in the 1950s, the US, Canada, Australia, and others enacted laws that required all food aid they donated to be "tied," i.e. produced by their own farmers and agricultural enterprises (Clapp 2012). The European Economic Community (EEC, now the EU) also began tying food aid as it built agricultural surpluses in the 1970s. These practices made sense for developed countries, to use up their agricultural surpluses and build new markets. They were, however, heavily criticized for undermining domestic production in developing countries with the effect of undermining both their food security and food sovereignty. Tied food aid was also expensive to ship from the exporting country and prone to spoilage in transit. Too much food aid or food that is not suitable for local markets in the wake of disasters will also rot in storage sheds or the holds of ships.

Developed countries, worried that US disposal of agricultural surplus as aid constituted dumping (overwhelming markets with cheap goods and driving out domestic and other external producers), pushed for international action. In what could be described as the first instance of global food waste governance, the UNFAO's Committee on Commodity Problems adopted the Principles of Surplus Disposal in 1954.

As agricultural surpluses declined in developed countries and criticisms mounted, they began to untie their aid, allowing instead grants, and purchase and distribution of food from the recipients' region or local area (known as local or regional purchase). The EU untied food aid in the 1990s and Australia and Canada followed suit in 2004 and 2005, untying half or most of their aid. The US is the only major developed country to maintain tied food aid. In 2018, agribusiness and shipping interests strongly opposed proposals to untie food aid from domestic sources in negotiations over the new US Farm Bill. Jennifer Clapp (2012) also notes that Japan, required to import a certain amount of rice annually under international trade rules, prefers to stockpile imported rice or divert it as tied food aid, as consumers prefer domestically produced rice.

Even with recent reforms, the practice of tied food aid casts a long shadow on domestic food production and markets in developing countries. Part of the point of tied aid was to create export markets once aid was no longer needed. It has changed tastes and preferences of local communities and disrupted long-standing agricultural systems. Developed country farm subsidies still boost exports that out-compete developing country exports on the global market.[9]

How this food is produced and culled or sorted before it ships has implications for local food security, sovereignty, nutrition, and cost. To cite one example, Havice and Reed (2012) find that in Papua New Guinea, tuna exporters sell on heads, tails, other offcuts or discards that are unfit for Northern markets to the local community. This practice undermines food sovereignty in subtle ways. It creates relationships of dependence because this resource used to be harvested by local fishers for local use and sale, not over-exploited by large foreign vessels for foreign consumption. Handing the offcuts and discards back to local markets certainly symbolizes relationships of dependency. To cite another example, exporters are often left with the imperfect produce, e.g. the green beans that are shorter or longer than the five inches required for the wealthy markets (Gille 2012). This has been an issue with Kenyan bean exports. These leftovers may flood local markets with culturally inappropriate (or non-nutritious) food.

Using this lens also brings into question prevailing definitions of food waste and loss. To return to the example cited in the UNFAO's official definition of food loss earlier in the chapter: those bananas that fell off the truck would not have been wasted in most parts of the world where bananas grow. There is no reason to assume they would have been left by the roadside; rather, they might well be gathered and consumed.

These contradictory, negative, or unintended outcomes of global "trade" in food and food aid provision, are an important but less well-studied piece of the global political economy of food waste and loss. They extend the North/South dimensions of food waste well beyond food waste in consumption versus loss in production. In this way, food waste (framed as agricultural surplus) crosses borders, connecting it to broader debates around the global politics of food and the structure of large-scale industrialized agricultural systems around the world.

Conclusions

The story of food waste and loss differs from that of electronic wastes. It is not as clear a narrative of global markets and resource extraction, nor is it such a flashpoint for debate and conflicting opinions. This case highlights the strength of activism and possibilities for innovative governance mechanisms at all levels, local to global, initiatives that have been put into practice. Food waste has high potential to be turned back into usable resources, through redistribution and reprocessing. Policies and practices such as changing date labels help avoid it arising in the first place.

Compared with the frustration and barriers that accompany many recycling efforts, effective composting of food scraps and waste may be a more achievable goal for municipal authorities. In a conversation I had with an employee of New York City sanitation department about China's crackdown on plastic and paper scrap imports and its devastating impacts on local recycling, she suggested that perhaps the only thing cities could do in the face of all this was to focus on composting biodegradable waste and do it well.

Like the case of discarded electronics, food waste highlights the importance of working at all points of a commodity's supply chain, from production to consumption, before it enters the waste stream. This chapter also highlights global implications of food waste for food security and sovereignty in developing countries. Food waste – shipped across borders or left behind when food is exported – is implicated in long-standing patterns of inequality in the global food economy. This global lens is often missing in conventional studies of food waste.

In chapter 6 we turn to look at the global political economy of plastic scrap and the high-profile nature of plastic trash entering the oceans. Food wastes and plastic wastes are connected through packaging. Plastic packaging helps keep food fresh and lowers the risk of contamination. Food wrapped in individual portions is less likely to be thrown away. Although alternatives are available, they are costlier and less convenient. The plastics industry is aware of this and touts its role in preventing food waste. However, when it comes down to it, much of the hardest to recycle plastic waste is packaging, especially soft films and plastic bags.

CHAPTER SIX

Plastic Scrap

In 2017 and 2018, public awareness of the global plastics crisis skyrocketed. Findings about the impact of plastics production and use revealed how much has been produced, how much has been discarded, and how long it lasts in the environment. Plastics travel through oceans, food chains, and into human blood and tissue. Momentum around the world to address this problem had been growing for years, but this period marked the most public attention paid to waste in a very long time. An article published in *Science Advances* in 2017 tracked the "fate of all plastics ever made" (Geyer et al. 2017). In December 2017 the UN declared plastics in the oceans a "planetary crisis." In June 2018, *National Geographic* ran an entire special issue entitled *Planet or Plastic?* Action crystalized in calls to ban or restrict single-use consumer plastics, such as straws and grocery bags. In May 2018, the EU proposed to ban some single-use plastic items such as straws and cutlery. By July 2018, even food retail behemoths like Starbucks were saying no to plastic straws. Also in 2018, the UN Environment Programme put out a report *Single-Use Plastics: A Roadmap for Sustainability,* urging action to curb single-use plastics.

Discussing discarded plastic as a global resource might therefore seem an unusual choice. However, plastics can be recovered, recycled, and reprocessed for manufacturing and energy generation. They are shipped around the world because there is demand for them, not just because generators need a place to dump them. They hold value under certain circumstances. But plastic scrap is a highly contingent resource: it degrades easily as it is recycled, and recycling processes pose health hazards to workers and communities.

So far, there are no immediate solutions to the challenges plastic scrap or waste pose. There is little sign of significant slowdown of plastics use and production anytime soon. Building markets and infrastructure for recycling and reprocessing plastics is badly needed. This case also shows the political vulnerability of the recycling economy, as decisions by a powerful actor – China, in this case – can disrupt this system completely.

In this chapter I argue that discarded plastic is a critical but contentious case in understanding the global political economy of waste. It straddles the border between waste in the traditional sense (has no value) and scrap (has value). Some urge we devote more energy to building recycling capacity and technologies to reclaim value. Others argue that our ultimate goal should be to reduce or eliminate our dependence on plastics rather than recycle or reprocess what we do produce. Circular economy initiatives at national and regional levels highlight plastics: their design and manufacture, their post-discard journey, and on minimizing or even preventing the use of plastics altogether. Although many agree that both paths need to be pursued, the tension between these two camps – recycle/reprocess versus eliminate plastics – is strong, and one of the main conflicts underlying competing visions of circular economies.

This chapter opens with a short account of our lengthy relationship with plastics, from love affair to disillusionment. It moves on to discuss how plastics disposal and recycling became global political problems, through ocean pollution and through the international scrap trade, becoming subjects for political debate from the UN down to municipal recycling authorities. It takes as its central case China's Operation National Sword, and its implications for plastics wastes and recycling, for marine pollution and for the as-yet undefined distinction between plastic scrap and waste at the global level. Is it even possible to extract value from plastic scrap or do we need to avoid it altogether? And what are the implications for trade in other forms of scrap?

Given that the global governance of plastic scrap and waste poses many challenges, we explore four options. The first is the worldwide rise of bans and restrictions on single-use consumer plastics, and the second is the quest for alternatives and substitutes for plastics in a range of applications. Then we turn to international law. The third option that has been put on the table is an international plastics treaty linked to ocean governance. The fourth is to regulate the trade itself, through existing international law or private sector initiatives. The 1989 Basel Convention on the Control of Transboundary Movements of Hazardous Wastes and Their Disposal is one, another is the World Trade Organization. Although the common assumption is that one would restrict the trade and the other facilitate it, experience to date suggests the outcome could be the opposite.

The Global Reach of Plastics and Plastic Waste

Modern society has had a long and complicated love affair with plastics, as Jeffrey Meikle's *American Plastic: A Cultural History* (1995), Susan Freinkel's 2011 book, *Plastics: A Toxic Love Story* and Rebecca Altman's 2015 essay "American Petro-Topia" recount.[1] From industry and manufacturing

to Tupperware and toys, plastics have pervaded the economy and everyday life. They have made life easier and
brighter for a long time. Plastic products are light, cheap,
easy to assemble, colorful, and disposable.

In 1950 global production of resin plastics was 2 million tons. By 2015, it reached nearly 380 million tons
(Geyer et al. 2015, p. 2). Plastics are produced by chemicals
corporations – Dow and DuPont (who merged in 2017),
LyondellBassell (the second biggest manufacturer in the
world, based across the Netherlands, the US, and the UK),
and BASF – and by oil companies like Exxon Mobil. Their
biggest use within the US by far is for packaging (34 percent
in 2017), followed by consumer and institutional goods (20
percent), and construction (17 percent), according to the
American Chemistry Council.

As time has gone on, it has become abundantly clear
that we produce and consume too much plastic, and
that existing disposal methods are entirely inadequate.
Further, plastics impose environmental damage and
health risks along their supply chains, from manufacture
to disposal. The petrochemicals industry is a major source
of pollution where oil is extracted and plastics are made,
becoming a focal point for environmental justice activism. Plastics are made from petrochemicals, which are
also the basis for beauty products, fertilizer, and other
products. By 2030 it is estimated that petrochemicals will
make up nearly a third of global oil demand, and half by
2050, according to a report from the International Energy
Agency (2018).

Direct health concerns from the use of plastics started
coming to the fore in the 1990s. Studies raised concerns
about chemicals used in plastic manufacturing that have
pervasive impacts on human health (Knoblauch 2009).
Phthalates (used to keep plastics flexible), bisphenol A

(BPA; used in water bottles), and others have potentially long-term impacts on human and animal reproductive systems and brain development. Almost all of us have measurable levels of phthalates and BPA in our bodies.

Other factors that have ended this long affair have to do with plastic waste, reflecting growing concern about the disposability culture and its permanent impacts on the environment. Engineered and industrial plastics have a service life of up to 35 years. Others, however, have a service life that can be measured in minutes, maybe seconds. No matter their service life, plastic products take anywhere from 5 to 1,000 years to break down, and even then, those microplastics may in effect last forever. Like JRR Tolkien's One Ring in his classic trilogy *The Lord of the Rings*, plastics are permanently destroyed only by incineration at extremely high temperatures (releasing toxic emissions into the atmosphere if they are not captured).

Geyer et al. (2017) estimate that 302 million tons of plastic waste were generated worldwide in 2015 alone, and 6,300 million tons cumulatively between 1950 and 2015. Of that, only 800 million tons has been incinerated and 600 million tons recycled – and only 10 percent of that has been recycled more than once. The rest is piling up on land (open ground and landfills) and in waterways, the main way plastics get into the oceans.

Two events shaped the globalization of plastic waste in the 2000s and 2010s. The first was the amount of plastic waste and debris found in the oceans and the impacts they have had on marine life, which came to light in the 2000s. The second, more recent, event was China's 2018 decision to effectively stop imports of plastic and other kinds of scrap, creating a shockwave cascading throughout recycling systems around the world.

Ocean Plastics

Nearly everyone has become familiar with sad and horr-
ifying images of marine life – turtles, seabirds, whales
– tangled in plastic or dead from ingesting straws, bottle
lids, and other pieces of plastic trash that have wound up
in the oceans. In 2018, the "Great Garbage Patch" in the
Pacific Ocean was estimated to be twice the size of Texas or
three times the size of France (Lebreton et al. 2018). Two
other huge gyres are in the North Atlantic and the South
Pacific, and these are only the largest of several. This case
shows the extent to which plastics have circulated around
the planet and into the global commons, affecting the health
of oceans and marine life (on top of climate change, ocean
acidification, and over-fishing), and ultimately humans.

Jenna Jambeck et al. (2015) estimate that 4.8 to 12.7
million tons of plastics have entered the ocean, including
plastic bags, bottles, caps, microbeads from beauty products, and nylon or plastic fishing nets. Discarded fishing
nets are estimated to make up 46 percent of ocean plastics
by weight. Discarded straws in their billions make up 0.3
percent of the total (Minter 2018; see also Dauvergne 2018).
This is lightweight in comparison to other oceans plastics,
but the sheer number of straws is harmful to marine life.
Ocean clean-up ventures can only do so much, but networks of ocean NGOs, such as the Oceans Conservancy, are
taking a lead in studying the problem and identifying and
implementing ocean-based solutions. Other organizations,
such as investment management firm Circulate Capital, are
looking for land-based solutions, notably improving plastic
waste management in coastal countries in Asia funded by
Northern donors and corporations.

Waste plastics in the ocean are grouped into macro- and
microplastics. Macroplastics are the large, identifiable

pieces – laundry detergent bottles, for example, or plastic bags – that wash up on beaches and float in territorial and open seas.

Microplastics are even more of a threat to marine eco-systems and food chains (Sundt et al. 2014). Microplastics include those that are intentionally made – like microbe-ads in cosmetic products or commercial abrasives – and particles that are byproducts or unintentional. Much of the Pacific Garbage Patch is made up of tiny pellets created as plastics degrade and fragment in the ocean. One significant type of microplastics in the ocean is tire dust. Microscopic fibers from laundering artificial fabrics on land are another source of ocean plastics. Such fibers are increasingly considered the main source of micro-plastics as work is done on sources beyond laundering, such as manufacturing. The Ellen MacArthur Foundation produced an influential report in 2016, *The New Plastics Economy Report* (along with the World Economic Forum and McKinsey & Co.). It predicted there would be more plastic than fish in the oceans by 2050 unless we change course away from business-as-usual (also because of the impacts of over-fishing, climate change, and ocean acidification on fish populations).

Most macroplastics flow into the ocean from the coasts of South, East and Southeast Asia, fast-growing countries that are typically producing a lot more non-biodegradable solid waste – including plastics – than they used to, without the infrastructure to dispose of and recycle them (Jambeck et al. 2015). Microplastics make up the bulk of plastic flows from industrialized countries. Studies show that plastic nanoparticles are absorbed by marine creatures at the lower end of the food chain, such as mollusks, and may work their way up the chain with uncertain impact on human food security and health (Barboza et al. 2018). A preliminary study found microplastics in human stool in subjects across eight countries (Schwabl et al. 2018).

Ocean plastics are a "global commons" problem: the high seas, like the atmosphere, are not exclusively owned by any country, and therefore are used by all, including for dumping garbage. Ocean pollution is also an issue of domestic politics in rapidly industrializing countries, which need to build management capacity for wastes generated while avoiding mass waste generation in the first place. Both dimensions of this problem create a need for global governance and intervention, particularly in the form of development aid and/or private sector investment.

China and the World's Plastic Scrap

Plastics are a recent entrant into the thriving global scrap trade (see chapter 4). In the late 1990s, with its economy booming, China became the leading global market. It started importing a large range of scrap, from plastic to steel, to meet the demand for inputs in its manufacturing sector. Plastic scrap had value, at least for a time. In March 2018, however, China stopped importing plastic, paper, and other types of low-grade scrap. Piles of discarded plastic quickly built up in ports and recycling facilities all over the world in the wake of China's crackdown. Municipal recycling programs in the US and other developed countries were devastated in the short term and were faced with difficult and expensive transitions in the medium to long term.

China's actions revealed how recycling has globalized, and how far wastes Western consumers send for recycling travel. They show how waste and recycling can get caught up in broader geopolitics and trade wars. The cross-scale implications of this case are especially fascinating: decisions made at the highest political levels in Beijing based on China's perception of its role in global politics directly affect what households worldwide put in their recycling bins every week.

This case shone a bright light into the grey areas of the global scrap trade. It has created major challenges for global governance: should discarded plastic be regulated under the World Trade Organization or under the Basel Convention? Given that the WTO enables global free trade and the Basel Convention aims to prevent environmental harm, one might assume that discarded plastics would be considered tradeable scrap under the former, and as hazardous waste restricted from trade under the latter. However, experience to date with both agreements suggests there is every chance the reverse could happen. We discuss this paradoxical possibility below.

Adjudicating between these two sides is, therefore, not easy. Trade and recycling of plastic scrap can help us out of the plastics crisis. Preventing this trade might, on the other hand, help wean us off an environmentally damaging product. However, following on from chapter 4, stopping this trade might be impossible and/or undesirable. For as long as we produce and use plastics, global circulation of scrap matters for building and maintaining a global circular economy.

The Challenges of Plastics Recycling

Plastics recycling at a municipal level began with the introduction of recycling programs around the developed world in the early 1980s. Since the early 2000s, according again to Geyer et al. (2017, p. 3), rates have slowly started to rise, but not enough to keep pace with the growth in plastics manufacturing. According to EPA data only 9.5 percent of discarded plastic that enters the waste stream in the US is recycled.[2] Of the plastic that is diverted for recycling, it is not clear how much of that gets recycled as much contaminated plastic waste is discarded at sorting facilities. And close to half of the rest went to China. Of the remainder, 15 percent is incinerated, leaving 75 percent to go to landfill. Europe's plastic recycling rate is just under 40 percent; China's was reported to be 29 percent in 2012 (O'Neill 2018).[3]

Why are plastics so hard to recycle in the US and other industrialized countries? The answers have to do with the different types of plastics, the few facilities in these countries to recycle them, and finding a market for reprocessed plastics (on the "violent afterlife of a recycled plastic bottle," see Winter 2015; see also Hopewell et al. 2009). First, recyclability varies across the different types of plastics. There are seven categories of plastic distinguished by their resin identification codes (RICs), stamped on the base of many plastic containers (see table 6.1). RIC numbers 1 and 2 are straightforward to recycle, creating sturdy and widely used products. The remaining plastics are hard to recycle efficiently or economically, and downgrade quickly with repeat recycling.

Second, many plastics sent for recycling are contaminated. Individual items may be dirty, such as plastic takeout containers still holding food, or unwashed milk jugs. Materials recovery facilities (MRFs, pronounced *murfs*) in the US have had to throw away up to half of what they are sent due to contamination. Bales of RICs 1 and 2 may be contaminated by other kinds of plastics and other items. Waste and recycling collectors, such as San Francisco-based Recology, point toward a lack of consumer education in terms of dirty recyclables, but also businesses and other larger entities are responsible for mixed plastics.

Table 6.1: Plastic types, uses and recyclability			
RIC	Name and abbreviation	Sample uses	Recyclability
1	Polyethylene terephthalate (mylar) (PET)	Food packaging, single-use beverage bottles, weather balloons	Easily recycled. Recycled into fleece, fabrics, tote bags, containers
2	High-density polyethylene (HDPE)	Milk jugs, garbage containers, and machine parts	Easily recycled into other hard plastics: detergent bottles, pens, drainage pipe, fencing, doghouses, floor tile
3	Polyvinyl chloride (vinyl) (PVC)	Construction materials, (pipes and siding); waterproof fabrics	Very hard to recycle. Contains chlorine. Can be recycled into lumber.
4	Low-density polyethylene (LDPE)	Plastic bags, food containers and six pack rings; plastic wrap	Not easily recyclable
5	Polypropylene (PP)	Plastic bottles, yogurt containers, rope, flip-top lids, straws	Not easily recyclable
6	Polystyrene (PS)	Packing peanuts, coffee cups	Not recyclable. Also contain CFCs.
7 (all other plastics)	Polycarbonate, acrylic fibers, polyvinyl acetate, synthetic rubber, Kevlar	CDs, rugs, blankets, emulsion in paints, tires, bullet-proof vests, brake pads	Not usually recycled

Sources: Steven Lower, *States of Matter*, chapter 9 at http://www.chem1.com/acad/webtext/states/index.html, Howard 2008 (a *Good Housekeeping* article), and *National Geographic's* Plastics Issue, June 2018. RIC = resin identification code

Third, it is hard to find uses (and therefore buyers) in the US and other developed countries for recycled plastic pellets and other products of the recycling process. High-tech plastics recycling is costly relative to prices obtained. There are also not enough facilities in many developed countries that can recycle plastics, especially the RIC 3s–7s. The US has built few if any recycling plants since the early 2000s, when exports to economically booming, low-wage countries started to take off.

These factors explain why there is a large supply of plastic scrap for export to countries where recycling and disposal is less expensive, therefore generating more revenue. They do not, however, explain the demand for plastic scrap in importing countries. To understand this, we turn to China's role as a scrap importer.

China and the Scrap Trade

In the early 2000s, China became the world's leading importer of scrap of all kinds, from high-quality industrial copper and aluminum to mixed bales of postconsumer plastic bottles, bags and containers. In 2016, China accounted for 27 percent of all global waste and scrap imports, including 55 percent of the world's copper scrap, 24 percent of its aluminum, 55 percent of its paper scrap, and 51 percent of global plastic scrap.[4] That year, 1,500 shipping containers laden with scrap of all kinds were put onto container ships for the voyage to China every *day* in the US (Flower 2016).

The trade made economic sense on all sides. China was not producing enough virgin plastic to meet demand in its booming manufacturing centers on its east coast. Its GDP growth rate reached a height of 14.23 percent in 2007 and averaged 10 percent in the 1990s and 2000s.

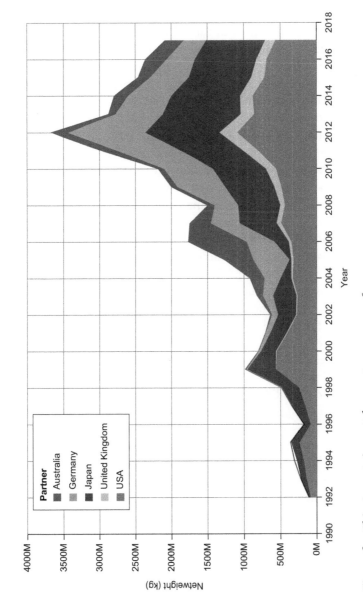

Figure 6.1: China's waste imports by country, 2007–2016

Source: Data from UN Comtrade

Even when China's growth rate slowed, its demand stayed high. A lower growth rate of 6.8 percent (2017) is still a large figure in absolute terms given the size of China's economy.

Competition for scrap among enterprises in China drove prices up, benefiting scrap producers. Shipping was cheap. From California, scrap was sent back in empty hulls of container ships that had brought goods over to the US, a process known as reverse haulage. For US-based waste collectors (especially on the west coast), selling scrap to a broker to be shipped to China was cheaper than shipping to a recycling facility or paying landfill fees.

What happened to imported plastics once inside China is not fully clear, but a considerable proportion was indeed recycled and reused according to industry reports (Velis 2014; see note 3). The labor and environmental conditions under which this happens are unknown, and it is also unclear how much may have been diverted to subpar facilities, including incinerators. However, China does have facilities, as it does with e-waste processors, that have reached comparable standards to more developed countries (Minter 2013).

So, why did China end this lucrative import business?

"No More Foreign Garbage"
On July 27, 2017, the Chinese government notified the World Trade Organization that it would crack down on imports of 24 types of scrap, most notably plastics, paper, and textiles.[5] By imposing a contamination limit that was almost impossible to meet, this action amounted to an effective ban (O'Neill 2017a). This decision sent immediate shockwaves through the international scrap industry. ISRI warned that this action, in the US alone, would lead to thousands of lost jobs, the closure of many recycling

facilities, and a sharp increase in the amount of waste sent to landfills.

China had cracked down on imports several times prior to 2018. In 2013 Operation Green Fence sharply increased inspections of imported bales, shipping back substandard ones at exporters' expense, making them pay more attention to bale quality. Immediately scrap started to get diverted to other ports, for cleaning or disposal. In 2008, Malaysia imported less than 20 million kilograms of plastic waste. In 2016, it imported over five times that amount. In 2016, US plastic scrap was exported to 78 jurisdictions including uninhabited islands.[6]

"Operation National Sword" (sometimes translated as "Sharp Sword at the Gate of the Country") began in early 2017, with port authorities inspecting about 70 percent of incoming scrap shipments. Rumors flew in the early months of 2017 that China was planning an even more substantive crackdown, but the extent of material covered, the stringency of the new standards, and China's rapid implementation timeline caught the global scrap industry by surprise. The new restrictions reduced allowable contamination rates in bales of paper and plastic from 3 percent to 0.5 percent. On March 1, 2018, despite a series of objections lodged by scrap industry and political representatives, China's decision went into effect.

This decision came from the highest levels of China's national government. Beijing had good reason to clamp down on plastic (and paper) scrap imports. Not only were shipments starting to overwhelm ports and facilities, they often exceeded existing contamination limits, particularly bales of paper and plastic. As well as being harder to deal with, they sell for lower prices, reducing tax and other revenues for port and local authorities. Scrap bales are also used to smuggle contraband drugs, electronics, and other banned items.

China's Ministry for Environmental Protection used the phrase "no more foreign garbage" when they issued their filing to the WTO. This phrase in and of itself reveals more complex motives. China is taking on a new role on the global stage. Already an economic superpower, through actions like taking a strong position on reducing greenhouse gas emissions and founding its own development bank, China also looks to be a leader by example. In this way, it seeks to shake off its high-profile image as the world's dumpsite. Beijing is working to replace China's informal recycling sector with cleaner, high-tech "eco-industrial parks," as we saw in chapter 4, with e-waste recycling in the city of Guiyu. These concerns dovetail with efforts to address China's broader environmental crisis, which has generated considerable domestic unrest with protests and lawsuits launched by China's growing middle classes (Stern 2013).

Specifically, noted filmmaker Wang Jiuliang spotlighted the plastic scrap trade in an award-winning 2016 documentary, "Plastic China," which focuses on an 11-year-old girl who lives and works with her family in a plastic recycling workshop. Amid the piles of plastics are many with labels recognizable to us in Europe and North America. Through interviewing the girl's family and the owner and his family (who also live on-site) and following their day-to-day work, the film details the realities of informal plastics recycling and its toll on their health and economic opportunities (at one point the owner buys a new car). The film went viral online in China after its release, then was quickly deleted from China's internet. There is speculation that this documentary, seen by high-ranking members of the Politburo, played a role in their decision.

Yet, there is a paradox here. Beijing wants to replace

imported scrap with domestically generated plastic scrap in manufacturing. As this plastic is reported to be of lower quality than imported scrap, this decision could worsen, not improve, environmental quality (Minter 2017). China does not yet produce enough plastic scrap to meet demand, either, therefore increasing its demand for virgin plastic resin. Local authorities also oppose these measures. They are losing even more revenue than they did with contaminated bales with no new scrap coming in. This opposition could encourage an under-the-radar trade through Hong Kong or across China's land borders.

From Scrap to Waste, at the Stroke of a Pen
The consequences of China's action for scrap exporters were quickly apparent, even before restrictions went into effect. By December 2017, many US municipalities, especially on the west coast, were badly affected as scrap continued to pile up with no place to go. By August 2018, nearly all US states felt the impacts.[7] Most notably, the UK, Australia, and Japan found their recycling capacity stretched to the limit (Laville 2017). Plastic and paper scrap once fetched AU$125–$325 per ton but, as of July 2018, were practically worthless, according to an analysis of the impacts of China's policies on packaging markets in Australia (Australian Packaging Covenant Organisation 2018). Affected municipalities began restricting the sorts of plastics and paper they allowed to be collected for recycling. Recycling collection companies started to raise prices. These higher prices were passed along to customers. In the US, states on the west coast were affected first, but impacts had filtered east by summer 2018.

Operation National Sword quickly reshaped cross-border flows of scrap that underpin recycling markets worldwide. In the wake of Operation Green Fence, other ports in

Southeast Asia were already taking in plastics for cleaning prior to shipping to China. Vietnam, Thailand, Indonesia, and Malaysia started taking in scrap in 2018 that would have gone to China. Brokers and traders in the US and other countries actively sought out other markets to take in plastic and paper.

Within a few months, the new markets – Vietnam, Thailand, and Malaysia – found themselves swamped. Thailand temporarily stopped plastic scrap imports in June 2018, enacting a permanent ban (that also included e-waste) in August. Vietnam and Malaysia stopped issuing or revoked import permits at the same time. Container ship backlogs at ports created headaches, and the possibility of illegal dumping of their contents into seas already receiving thousands of tons of land-sourced debris.

Finally, this story unfolded during a trade war between the US and China in 2018. As President Trump levied punitive tariffs on Chinese imports, the Chinese government retaliated by imposing 25 percent tariffs on $16 billion worth of US imports in August 2018. This package included many types of scrap beyond the 24 named in 2017, such as ferrous scrap, nickel, copper, zinc, tin, and aluminum – in other words, the most valuable categories (Rosengren 2018b).

China's actions revealed the extent to which the high-consuming nations had become dependent on the high-growth nations for their waste disposal. They compelled industrialized countries and their local and state-level governments, cities, and citizens to rethink their dependence on overseas disposal for recyclable trash. Many developed countries had simply outsourced their recycling rather than build new facilities, especially for lower-quality plastics and paper. Whether or not these restrictions are permanent, governments in US states and other countries had to plan new recycling infrastructure.

In the US, state and municipal governments worked to upgrade MRF facilities for both plastic and paper. Such measures included plans to reopen long-closed pulp and paper mills. In early 2018, the New South Wales state government in Australia announced an AU$47 million short-term support package for local recycling authorities in the wake of Operation National Sword. Circulate Capital, an organization spun off the Closed Loop Foundation in 2018, was set up to build waste disposal infrastructure in Southeast Asia, driven by the amount of ocean plastic originating from these countries, but also to combat the impacts of Operation National Sword. Finally, Chinese companies have visited the US, planning to build recycling facilities there.

One study has estimated that Operation National Sword could displace an estimated 111 million tons of discarded plastic by 2030, concluding that "bold global ideas and actions" for boosting recycling capacity, reducing dependence on plastics and redesigning products are sorely needed (Brooks et al. 2018, p. 1). The next section examines how global – and local – governance initiatives are responding to this challenge.

New Directions for Global Plastics Governance

The plastics waste crisis poses challenges for global governance, but also opportunities for innovation all along the plastics supply chain, from production to final disposal. Plastics are everywhere, with an infinite number of sources, making governing their production and use worldwide hard. It is difficult to intervene in national and corporate decisions, especially when plastics manufacturing corporations are among the most powerful on the planet. It is easier for global actors to intervene when wastes and scrap cross national borders or enter the oceans.

The impacts of Operation National Sword facilitated the push to reduce our dependence on plastics triggered by the ocean plastic crisis. Zero Waste Europe, GAIA and other zero waste organizations have launched campaigns that make these connections. #Breakfreefromplastics is a transnational activist alliance that coordinates efforts by groups and individuals across the world to control plastics pollution and end plastics consumption. These campaigns have visibly influenced consumer choices and new legislation and initiatives to cut down on single-use plastics. They have advocated a global legal agreement on plastics, whether as part of an existing treaty or as a stand-alone entity. The scrap industry is also pushing for global solutions, whether through existing international law and/or the development of verifiable standards that are acceptable to China and other markets.

These efforts reveal different dimensions of global environmental governance and how it is being done, which may also have relevance for other arenas of global governance, including global climate governance. So far, rules and guidelines around plastic waste enacted around the world have focused on single-use consumer plastics. In 2018, international organizations, governments and advocacy groups started taking the prospect of a new treaty on oceans plastics seriously. Highly contentious global politics also revolve around the trade in plastic scrap: should it be allowed or not? Is plastic scrap a dangerous waste, or one to be traded, at least in some form? These proposals look to existing trade governance mechanisms: the Basel Convention and the WTO.

Single-Use Consumer Plastics

Many efforts to reduce plastic waste around the world involve cutting back or eliminating single-use plastic products, including lightweight shopping bags, single-serve water bottles and plastic packaging. In 2015 and 2016, the single-use straw joined this group of "battleground" products in the fight to reduce consumer plastic waste (Toscano 2018).[8]

Single-use lightweight shopping bags were one of the first targets. Early plastic bag restrictions were implemented in South Asia (Clapp and Swanston 2009). Bangladesh declared a ban in 1998 following an incident where bags blocked flood drains for two months, preventing water levels from falling. During the late 1990s, many Indian states and jurisdictions also legislated against bags. Taiwan began phasing out bags in 2002, and South Africa implemented a levy in 2004.[9] Full bans exist in several African nations, and local bans are in place across Latin America. Kenya's measures to end "flying toilets" received a lot of attention, including as they did prison terms and fines of up to $40,000 for violators. In their first year, they achieved considerable success in reducing litter and pollution (Watts 2018). While some of these measures have been rolled back or faced difficulties in implementation, they predate measures in the global North by some years and marked an emerging global norm against lightweight plastic bags.

EU member countries primarily levy customer charges on plastic bags, although regulatory restrictions will be in place by 2030. The EU has targeted ten single-use plastics and fishing gear for phaseouts over the near to medium term. These measures are not always implemented smoothly. In July 2018, the Australian supermarket chain Coles introduced a bag ban, then backflipped after customer complaints, and counter-flipped after more complaints – all within weeks. As of 2018, only California and Hawaii in the US had implemented bans, in the face of heavy resistance from the chemicals industry and bag producers, although levies and restrictions exist in many other states, and bans at county and city levels.

Industry has mobilized to fight back against bag measures, notably in the US, directly opposing legislation, blaming littering by individuals, and backing preemptive legislation. The American Progressive Bag Alliance, supported by many plastics manufacturers, mounted a campaign to reverse a bag ban in California in 2015. At least eight US states have adopted legislation to preempt local authorities implementing restrictions on plastic bags (Florida, Wisconsin, Indiana, Iowa, Michigan, Mississippi, Missouri, and Arizona).[10]

The global spread of bag measures illustrates one form of global governance, namely transnational policy diffusion, where policy measures developed and implemented in one country or jurisdiction are picked up by another. Studies have started to show that plastic bag measures, which have been in place for longer than other consumer waste restrictions are having positive impacts. One study (Maes et al. 2018) found a significant reduction in plastic bags on ocean floors between 1992 and 2017 (although an increase in other categories, such as fishing gear). The report suggests that consumer product restrictions and changing cultural norms have long-term effects.

A movement to ban or restrict the use of plastic straws gained global momentum in 2015. Billions of straws are discarded around the world each year.[11] They are a highly visible example of disposable consumption, with a use time, often, of seconds. By 2018, large corporations, from

Starbucks to Caribbean Cruises to the Marriott hotel chain, announced they were giving up plastic straws. The UK announced a ban, as did some cities, such as Santa Barbara, California. In 2016, the town of Blackheath in Australia's Blue Mountains may have been the first town in the world where businesses phased out plastic straws in favor of paper. Targeting plastic straws is a symbolic gesture. Some say it is futile: that it misses the larger problems such as discarded fishing gear, or that it distracts from broader structural issues, such as the political clout of the plastics industry. Its power, however, comes from the way it highlights an action many consumers barely pay attention to, connecting it to broad and serious environmental impacts.

Large corporate users of plastics, including Coca Cola, Walmart, and Unilever, are looking for ways to cut back their overall use. They are taking back used bottles and packaging, developing new containers, such as thinner plastic water bottles and donating to ocean cleanup and waste infrastructure projects in coastal countries in Asia. Over the longer term, however, the race is on to find alternatives and substitutes for petroleum-based plastics that can be produced in large quantities at reasonable cost and lower environmental impact. This is proving a difficult quest, but one that many have taken on.

Substitutes and Alternatives to Fossil Fuel-Based Plastics

Over the longer term, activists, policy-makers, and industry advocate for a range of alternatives for addressing plastic waste, and many see Operation National Sword as an opportunity, not a disaster. These alternatives include phasing out the use of conventional plastics, incentivizing better (e.g. compostable) plastics, and developing plastics substitutes and alternatives.

Developing substitutes (non-petroleum-based plastics) is not simple. The main substitutes for conventional plastics are bio-based plastics, made, for example, out of cornstarch, straw, or even food waste (Spierling et al. 2018). They are designed to be fully biodegradable in the environment. In fact, compostable cutlery and plates need specialized facilities to be fully broken down.[12] Plastic packaging is hard to replace with substitutes that offer the same level of hygiene, resilience, and maintained freshness. Many retailers worry that bio-based plastics are not as able to tolerate heat. In fact, the plastics manufacturing industry has advocated the role of conventional plastics in preventing food waste. Alternatives to plastics include paper, cloth, and other alternative materials for packaging and other uses (including stainless steel straws). Reusable coffee mugs and water bottles are now ubiquitous around college campuses in the US. These are often costly. Paper straws cost around four times as much as plastic straws, which places burdens on the retailer or the customer. In Kenya, a journalist interviewed a market vendor who, after the bag ban, was complaining about the more expensive cloth alternatives she needed to use (Watts 2018).

There are other innovative plastics substitutes being developed on a small scale. Mycelium biofabrication (developed by Evocative, a New York-based advanced materials company) creates products designed to replace polystyrene foam, thermal plastics that can hold in cold or heat. "Designing out" plastics is a cornerstone of circular economy platforms (see the conclusion of the book).

Another alternative – waste-to-energy incineration – has again come to the fore. Zero Waste and anti-incineration advocates have heavily criticized initiatives such as the

Hefty Energybag Program, a pilot initiative in Omaha, Nebraska, to divert plastics to energy production. But small companies like Salt Lake City-based Renewlogy have worked to develop newer, cleaner technologies to convert plastics to energy, including thermal depolymerization. However, finding an affordable, scalable, environmentally-friendly, non-fossil fuel-based substance that can replace or imitate plastics remains a distant solution. This means that others are directing their energy to global politics, and the prospects of utilizing global governance mechanisms to address the plastics wastes crisis.

A Global Plastics Treaty?

In March 2018, the Sixth Annual Marine Debris Conference, meeting in San Diego, California, brought together oceans, plastics, and waste stakeholders. Attendees discussed, among other topics, how global plastics and oceans governance might be brought together in a new multilateral agreement that would bring countries together to address these problems.

The international community has long relied on multilateral agreements as a means of solving or managing shared environmental problems, often negotiated under the auspices of the United Nations (O'Neill 2017b). The best known is the 1992 UN Framework Convention on Climate Change and its component agreements: the 1997 Kyoto Protocol and the 2015 Paris Agreement. There are hundreds more, including a set of agreements that successfully phased out or drastically reduced the production and use of ozone-depleting substances: the 1987 Montreal Protocol and subsequent amendments is widely considered one of the most successful global environmental governance regimes.

Human activity has threatened the world's oceans in many ways, including over-fishing, ocean dumping, oil pollution and ocean acidification, and there is a large body of international law designed to protect them. Now the flood of plastic into the world's oceans creates another threat (Dauvergne 2018). It can potentially be harnessed to existing agreements, for example as an amendment, protocol, or annex to the UN Convention on the Law of the Sea or as a free-standing agreement that connects these two issues.

But what would such an agreement look like, and how politically feasible would it be to draw up and implement? How willing would countries (and corporations) be to give up their manufacturing and use of plastics? Rather like climate change and fossil fuels, plastics production and use is so ubiquitous, it would be hard to mandate a phase-out. As we saw above, unlike with chlorofluorocarbons (CFCs; the main ozone layer-depleting substances), there is no easy substitute. Generally speaking, it is always difficult to draw up any international agreement, given countries' competing interests.

There are, however, tactics that advocates of a global plastics agreement can use. One is to frame an agreement to target a specific issue that everyone, from activists to plastics producers, would like to stop. In this case that issue is plastics entering the oceans, especially from the coasts of Asia and Southeast Asia, which would allow different actors to target waste management capacity building. Another tactic is to develop multi-stakeholder agreements, which would give NGO and corporate sector representatives a voice in decision making. Third, an agreement could contain binding and non-binding elements, building on the approach taken in the Paris Agreement, where countries set their own targets within a broader legal framework. Nearly 200 countries signed a non-binding UN Resolution in December 2017 to stop plastics entering the sea. Goal 14 of the Sustainable Development Goals, on the health of the oceans, is another venue for steering countries toward stronger outcomes.

Global Governance of the Plastics Trade: Scrap or Waste?

The governance approaches discussed above all focus on plastics' production, use, and disposal. There is another piece of this puzzle: what to do about the global plastic scrap trade. Although the trade itself represents a small portion of waste plastics overall, Operation National Sword, as already noted, could displace 111 million tons of plastic by 2030 (Brooks et al. 2018) and much of that is likely to wind up in landfills, in the air via incineration, or in the oceans. Under what conditions does plastic scrap hold value? Does it depend on type, condition, or the conditions under which they are treated in the importing country? As with the example of cardboard boxes at the start of chapter 1, one person's (or one country's) waste is another's scrap, only in this case we are talking about global markets worth millions of dollars.

Further, to connect directly to the themes of this book, what happens to the plastic scrap case has implications for the entire global scrap trade. Operation National Sword also cracked down on paper and other scrap, and China used scrap imports as a weapon in its trade war with the US in 2018. It is also highly relevant to the electronic waste trade. E-waste and plastic scrap are both marginal types of scrap, falling on the border between waste and scrap. They have negative impacts, but also contain extractable resources, whether through recycling, resource extraction or repair. But again, there is strong disagreement over whether trade should continue in these cases, creating uncertainty that could have ripple effects across the entire scrap trade.

.

There are three distinct options available for governing scrap trading at the global level. First are scrap certification initiatives. Voluntary certification and eco-labeling of products is a way private sector actors can signal that their products are environmentally (or socially) sound, and has been applied to timber trade, fisheries and other traded goods (see also chapter 4), as well as to social and labor standards garment manufacturing. ISRI established the Recycling Industry Operating Standard (RIOS) certification for exporting scrap in 2005. In this case, RIOS certification would mean scrap is prepared and cleaned to the standards of the importing country, in this case, China. Such measures are acceptable according to international trade rules. However, they may run into some of the same problems other certification systems have, for example maintaining verification standards. Critically, it is also hard to monitor and verify work and safety conditions within the importing country under such systems.

The other two options involve situating plastic scrap into one or other of two sets of trade agreements: the Basel Convention and the WTO. Neither option has an obvious outcome: for complicated reasons, either could lead to ending or to facilitating the plastic scrap trade.

The main international venue for regulating international trade waste is the 1989 Basel Convention on the Control of Transboundary Movements of Hazardous Waste and Their Disposal (see also chapter 4). Set up initially to obligate exporters to obtain prior written consent from potential importers, in 1995, member countries adopted an amendment to the convention that would ban waste trading from OECD to non-OECD countries. The so-called Basel Ban has not yet been ratified by enough countries to enter into full legal force (after over 20 years). Instead, it has shaped the hazardous waste trade by creating a normative consensus against transboundary dumping and encouraging monitoring and reporting of possible cases.

In June 2018, Norway proposed adding plastic waste to Annex II of the Convention, to the category of household wastes of special concern. Among other things, treaty annexes list technical details too long for the main text. Annex II states that wastes collected from households and residues arising from the incineration of household wastes should be subject to "special consideration," meaning they can only be exported from an OECD to a non-OECD country with the prior written consent of the importing country.[13] This could lead to restrictions on plastic scrap exports from Northern to Southern countries (including, in this case, China, categorized as part of the Global South under Basel). Banning this trade means reducing negative impacts caused by plastics pollution to workers, land environments and the oceans, and shifts the burden back to the consuming country. It may (just may) have the effect of incentivizing innovations, but only if it can be enforced, which, in the case of electronics, has not been the case. However, it will also shut down foreign markets where the scrap will be reprocessed and reused, increasing waste dumping and, potentially, transboundary smuggling.

Norway's proposal was discussed in a Basel Convention Working Group meeting in September 2018. While it had strong support from developing countries and China, others, including the EU, opposed the move. There are a number of problems with taking this route. First, it is not clear if plastic meets the required conditions to be defined as hazardous under Basel, in that it is not hazardous in and of itself, although this is open to interpretation. Also, importers or exporters can make their own decisions. Second, it could undermine support for the Basel Convention and Ban if countries with an interest in the plastic scrap trade view it as over-reach. It would also only restrict North-to-South trade, while as we saw with the e-waste trade, the growing problem has been trade among the countries of the global South – in this case, all non-OECD countries.

There are two ways the WTO could become a significant venue for regulating the plastic scrap trade. First, one of the biggest concerns for the Basel Ban is that its enforcement could trigger a trade dispute at the WTO should it enter into force as it is written. The WTO governs international trade. It and its predecessor, the General Agreement on Tariffs and Trade (GATT) have pushed for trade liberalization since the late 1940s, encouraging easy global movement of goods and services.[14] Member countries who believe another state has enacted restrictive trade measures can bring a dispute to the WTO for arbitration. If a country such as India (as it once threatened to do) claimed unfair trade restrictions, or discrimination, because it had been arbitrarily excluded from a list of countries able to import recyclable wastes (as it has been) under the Basel Ban, it could win.

But this is only one way the WTO could get involved. There is another potentially more significant avenue that could yield entirely unexpected results: the adjudication of trade-related disputes directly between countries. Since 1995, over 500 cases have been brought by countries to WTO dispute settlement panels.[15] A small but significant minority of those have been directly related to environmental measures imposed in one country that another believes discriminates against its goods. The outcomes of these cases are complex. They do not reflect a straightforward victory of economic globalization over environmental protection. However, it is the main global venue where China could (eventually) be compelled to resume scrap imports.

The only WTO dispute so far that has directly addressed waste/scrap exports and imports was launched by the

EU in 2006. In the WTO–Brazil Retreaded Tire Case, the European Commission (the governing body of the European Union) filed a complaint against Brazil for refusing to import retreaded (used) tires.[16] The EU claimed Brazil was protecting its own tire market. Brazil claimed it was left to deal with end-of-life tires and the hazards they pose, from fires to becoming breeding grounds for disease-bearing mosquitoes. Although Brazil technically lost the case in 2007 on the grounds of discrimination against imports from a single trading partner, the WTO allowed it to restrict *all* imports of used tires on health and safety grounds. This case, while obscure, may hold clues for future governance of the scrap – and waste – trades.

Getting such a case to the WTO and subsequent deliberations would be a complex and lengthy process, but it may be the best venue to harmonize a global definition of scrap for trade purposes. As the EU–Brazil case shows, these decisions do not always rule against environmental protection (in this case, recognizing that retreaded tires caused environmental and health damage in Brazil), but may keep the door open to regulating scrap as a tradeable good, not solely as an environmental hazard.

Conclusions

> Recycling doesn't work because of technology, values, or intentions. It requires strong and stable markets for scrap and recycled goods (paraphrased from Minter 2013).

Plastics are not the instantaneous cataclysm of nuclear war, nor the slow rolling violence of climate change. We are uncertain of their long-term effects, but, in their stealthy ubiquity, they represent yet another dimension of humans' deep impression on the planet.

This chapter was written as the developments mentioned

in it were unfolding. Plastics and plastic waste or scrap burst onto the global political scene unexpectedly and with considerable momentum, driven by awareness of the ocean plastics crisis and China's Operation National Sword in the mid-2010s. The global scrap trade has a longer history, but it too has been disrupted and thrown into the limelight in the wake of National Sword.

As far as plastic scrap goes, we are on a very shaky global resource frontier. It is contentious to even talk about plastic as scrap. Indeed, in some of the analysis of Operation National Sword, there is an element of deserved retribution, that the wealthy countries are finally being punished for their shiftless ways. Yet, it is not clear that global trade in plastic waste, as with used electronics (chapter 4), can or possibly should be stopped entirely. There are few markets, at least for now, in the global North, for plastic pellets and other recycling outputs, markets that are necessary for recycling to make economic sense. Therefore, plastic scrap may become subject to international trade rules as a good, or to international environmental rules as a hazard. What ultimately happens may depend on how these cases are brought up, and by whom: an aggrieved exporter, an aggrieved importer, or another outside party.

In thinking more generally about the future of plastics, the quest is for a larger solution that will facilitate the development of substitutes and alternatives for fossil fuel-based plastics of all kinds. Plastic has taken a front seat in many different circular economy plans, including that of the EU. The concluding chapter delves into a deeper discussion of the global circular economy, the vision of a post-waste future espoused by many organizations and policy-makers around the world, from Europe to China.

Conclusion:
A World without Waste?

This book has unpacked some of the many ways in which wastes are (or can become) resources, rather than stuff to be thrown away, out of sight and out of mind. In so doing, I have focused on the global waste and recycling economy: wastes' production, their transformation into resources that are bought and sold on global markets, and their final disposal within, across, and beyond national borders. This book began with three themes: wastes as a *global resource frontier, magnified risks* from a waste economy that has grown in scale and scope, and all the *governance challenges* that go along with managing wastes as a globally networked commodity, but one that has attendant risks and unequal impacts.

This "new" global waste economy emerged for several reasons. Humans produce, at ever increasing rates, too much waste of all types. We now understand more about how much of value is trapped in that waste (discarded electronics being one major example). And this economy has globalized: wastes move relatively easily to where they will be dismantled, recycled, reused, or dumped. Transnational networks of actors are engaged in these transactions, and tens of millions gain their livelihoods from waste work.

Understanding wastes as either risks (or externalities) or resources (or commodities) is not enough. Two additional perspectives help make sense of the global waste world:

wastes as livelihoods (capturing dimensions of waste work) and wastes as inputs (speaking to circular economy discourses where wastes are cycled back into production). I also mapped these six perspectives (see box 1.1) onto different sources of value of wastes – in collection and disposal, in recycling and resource extraction, and in prevention (see chapter 2). Finally, wastes are contingent resources. What counts as waste (or as resource) is highly context-dependent. The transformation of wastes may be uneven (not all components of a cell phone contain value), and their potential to become resources is highly dependent on market stability.

The case study chapters drew out these themes in all their complexity. Chapter 3, on waste work, dealt with the formal and informal waste economies and the intersections between them. Chapter 4 tackled discarded electronics, chapter 5, food waste, and chapter 6, plastic scrap. All chapters took a global perspective but also brought the analysis down to the local level (literally to street level in some instances).

Wastes as a Global Resource Frontier

Economic growth does not happen solely through the allocation of scarce resources. It is also driven by finding and exploiting new resources, namely, opening new resource frontiers. This has happened historically with land, gold, oil, timber, fish, and other resources, some of which are now close to exhaustion. Wastes are increasingly exploited for resources and energy. Studies have brought home how much material is lying around or under the planet's surface, just waiting to be mined. Perhaps as much as 30 trillion tons have accumulated since the Industrial Revolution (Zalasiewicz et al. 2017).

Wastes are an unusual resource frontier. First, they are not geographically isolated. They are everywhere, all around us, in every community. They can be easy to access or enclosed in a fortified facility. They are highly dispersed but also piled into "mega-dumps" that stretch for hectares. Wastes are often highly mobile. Unlike traditional ores, wastes can be shipped somewhere else entirely to be "mined," but they can be smelted and refined alongside traditional ore. These characteristics were captured in the example of the flexible mine for e-waste in chapter 4 (Knapp 2016).

Second, wastes are highly differentiated. To use Gourlay's example, the concept of waste can encompass a blob of mustard on the side of one's plate and a dismantled nuclear reactor (Gourlay 1992, p. 20). What brings them together is that they are "what we do not want or fail to use" (Gourlay 1992, p. 21), yet we still need to know subsequently how to use or exploit them. This needs to be seen on a case-by-case basis. Food waste, used electronics, and plastic scrap are all very different from each other in terms of the treatments they need, the amount of time they last in the environment, the sorts of hazards they pose, and the value they may contain. There are strong variations within those categories as well, for example some plastics (as identified by their RICs) are far more recyclable than others.

Nonetheless, many different types of wastes are recycled, reprocessed, refurbished, or reclaimed for public or private benefit. The potential profits from fully exploiting the value contained in wastes are sizable. The *Global E-Waste Monitor* reported that €55 billion of valuable materials are trapped in the world's reservoirs of e-waste (Baldé et al. 2017).

This book has shown how, and why, wastes are a *global* resource frontier. They are shipped all over the planet, as we saw with electronics, food, and plastic scrap, for recycling or disposal. They tend to be shipped not to where they will be dumped, but to where there is a market and manufacturing capacity to reprocess the scrap and reuse it. Local waste economies are embedded in the global. Communities of waste pickers dismantle electronics shipped from far away. Multinational mining companies engage in urban mining in Southern cities. The case chapters emphasized the need to look at demand for imported wastes, not just the supply, to understand the directions and persistence of waste trade.

As with all resource frontiers, there is conflict, often as multinational interests run up against local communities in order to capture the profits from exploiting this resource. Governments get involved, often to displace informal sector workers from the source of their livelihoods. The case of Cairo's Zabaleen (chapter 3) showed how Egyptian authorities attempted to repress this traditional recycling community and to allow multinational corporations to take over waste collection. The Chinese government has intervened to clean up its e-waste and plastics recycling economies, replacing the disorder of many informal facilities with orderly industrial parks.

Who benefits from the exploitation of wastes? The answer is not obvious. Collecting and selling waste and scrap have traditionally been paths to upward mobility (Zimring 2015). Even today, informal waste-picking pays more (and has lower barriers to entry) than other forms of informal work. At a global scale, some of the world's largest corporations – utility companies such as Veolia or corporations within the waste/recycling sector such as Waste Management – make money not so much from waste collection and disposal than from large-scale recycling and energy generation. Compost collectives and repair stores in cities and towns provide community benefits and perform essential services. Many small firms and innovators who are developing plastics substitutes are reaping value from waste prevention. This world is so diverse that profits do not always go to the most powerful.

Magnified Risks

With this globalized resource frontier, however, come magnified risks. More waste produced equals more opportunity for risky outcomes, from dangers to workers to increased greenhouse gas emissions from landfills packed with discarded food. Over the past decades, hazardous and nuclear waste have had to be sequestered in limited space, given their persistence in the environment. More and more plastic debris floods the oceans, threatening marine life and food chains. The concept of distancing – both literal and figurative – helps to conceptualize these globalized risks (Princen et al. 2002). As wastes are often disposed of far from points of use, they are both out of sight and out of mind for the average consumer, which has been an obstacle in gaining effective support for potential solutions.

There are also risks apparent for workers in the global waste economy. Informal workers dismantling old electronics with hand tools and little protection are exposed to mercury, dioxins, and other toxins. Informal waste workers are among the most vulnerable to disease, weather and other dangers, including landfill collapses or slides in the world's mega-landfills. The case chapters showed how workers are vulnerable to displacement from their livelihoods without compensation due to incursions by governments and corporate actors.

Markets for recycled and reprocessed materials are highly vulnerable to economic or political shocks. If primary material prices fall, secondary prices fall further. China's Operation National Sword demonstrated just how quickly global recycling markets could collapse, with knock-on effects on waste collection and management in the US and other exporting countries. It also showed how quickly global markets could lead to diversion of wastes to countries like Malaysia, Thailand, and Vietnam, who were not prepared for the influx. This example also shows the ability for these countries, considered weak in the global system, to resist unwanted imports, as each imposed their own restrictions within months of Operation National Sword going into effect.

Some proffered solutions to the global waste crisis are highly controversial. Waste-to-energy incineration is one. It has scaled up tremendously over the past 10 years in upper income countries, driven by China's embrace of the technology, from an 0.1 percent share of energy generation in 2012 to 10 percent in 2018 (Kaza et al. 2018, p. 2). Opposition from organizations such as GAIA has helped mobilize opinion against WTE incineration given its long legacy and ongoing problems of air, water, and ground pollution (Baptista 2018). Others argue that thermal WTE treatment has to be part of the repertoire of tools to deal with waste (and reduce greenhouse gas emissions), and that newer technologies offer more promise with less pollution (Castaldi 2014). This is an ongoing debate.

Governance Challenges, Opportunities, and Initiatives

Even before I started the research for this book, it was painfully obvious that existing mainstream modes of waste governance and regulation were inadequate for the new global waste economy, its opportunities and its risks. Domestic waste regulations based on wastes as something to be taken away and disposed of failed to capture products' afterlives, as, for example, with electronic wastes. Recycling collectors could just ship the collected electronics overseas for a quick buck, for themselves.

At the global level, international legal arrangements had been put in place at an earlier time for a different problem. The 1989 Basel Convention on hazardous waste trading addresses a very important global environmental justice issue – dumping of hazardous wastes from rich countries to poor countries. However, its structure is inadequate to oversee more complex cases where, as is the case with e-wastes, many shipments occur between Southern countries. Nor can it easily address substances that are not shipped as wastes (e.g. computers labeled for charity) or that are not directly hazardous in their current form (e.g. plastic scrap).

One of the positive findings of this book is just how many governance innovations and experiments are going on in the waste economy, from local to global, despite many remaining challenges and uncertainties. As argued in chapter 1, making the global resource frontier visible has enabled the creation of effective governance mechanisms that encourage the reuse of these valuable resources while mitigating magnified risks. Such governance initiatives are designed by national and local governments, private sector actors and NGOs, and global agencies.

Waste activism is strong and multifaceted. Waste picker organizations have networked across borders to represent shared concerns and magnify their voices. Food waste and plastic waste activists work in local supermarkets and restaurants. Large NGOs around the world have added waste campaigns to their work. Zero Waste NGOs such as GAIA form transnational networks, while students mobilize on college campuses.

Recognizing wastes as both streams (after disposal) and as objects (with an entire lifecycle) has broadened policy approaches. Food waste is a good example: advocates of waste prevention have designed initiatives that deal with food that has been discarded and also ones that recognize the importance of looking at food's entire lifecycle, from production to discard.

Many people concerned with the e-waste trade have recognized that efforts to stop the trade are likely to be futile, possibly counter-productive. Alternative governance initiatives therefore focus on how to manage it and ensure that risks are minimized, while benefits are shared with those who repair or dismantle used devices where they are shipped. Studies of the e-waste sector in Ghana have, for example, shown high rates of electronics repair and resale, not just a burning wasteland, signifying the existence of an "informal green economy" (Minter 2016).

In one case at least, governance initiatives are flailing, not yet finding good or acceptable solutions. Plastic scrap remains highly problematic, although energy levels around finding solutions are high. In other cases – such as pushing for the re-design of electronics to last longer, to be easier to repair, and to be safer to be dismantled – powerful corporations have so far resisted calls to change product design so fundamentally.

At global and regional levels, government actors are working to create comprehensive governance frameworks that take into account wastes' value as resources and the need to minimize their generation. The UN's Sustainable Development Goals (SDGs) have clearly articulated elements that address waste prevention, most notably setting specific goals around food waste. The one issue that looms over the global community right now is the question of the scrap trade: under what conditions are discarded plastics or electronics scrap and when are they waste? How this definition is worked out (if it ever is) will shape the global waste economy for decades to come.

Other Cross-Cutting Themes

Researching and writing this book led to additional insights. In addition to waste trading appearing in every case, and the vibrancy of transnational waste activism, I highlight the most important here.

First, this analysis *overturns conventional narratives about North and South* in the global waste economy. No longer can it be said that the countries of the global South are passive victims of waste dumping from the global North (something that may never have been fully the case; see Montgomery 1995). Markets exist in manufacturing centers in China, Ghana, Nigeria, and other countries that can reprocess and use these materials. Josh Lepawsky's work has done the most to reveal the dense and multi-directional networks of trade in electronic wastes (e.g. Lepawsky 2018).

Second, *product design matters*, as do entire supply chains, in minimizing volumes of wastes and their associated risks. Electronics contain valuable metals but also substances that are highly dangerous. Newer generations of electronics are more likely to be thrown out as manufacturers incorporate components (such as batteries) that cannot be removed or replaced. In turn this has ignited a *"Right to Repair"* movement that challenges companies' legal ability to do this or encourages the establishment of repair facilities.

Third, most of this book was written as China threatened to announce, announced, and then implemented *Operation National Sword*, between January 2017 through to September 2018, which closed its ports to plastic and paper scrap imports. I had known about the extent of

China's scrap imports since 2013, with the first crackdown (Operation Green Fence), and through supervising a UC Berkeley senior, whose thesis topic led her to trace where plastics put in campus recycling bins went (Davis 2014). I read *Waste Dive* and industry journals as rumors of something bigger started in early 2017, up to the July announcement, which came as a shock in terms of its severity and short timeframe. Subsequently, I have been on panels and in conference halls, hearing from waste industry representatives presenting their case to the Chinese government and people working at Northern Californian recycling facilities dealing with floor to ceiling piles of old paper. It has been a big deal and continues to unfold. From my perspective as a global (environmental) politics scholar, two factors stand out. First is the extent to which geopolitics plays a role, along with the changing role of China in the global economy. Second, this case illuminates how events and decisions play out *across scales*, from the Chinese national government and the WTO right down to curbside recycling collection practices and policies in towns and cities across the global North. It illustrates extremely well how local and global waste economies are inextricably intertwined.

We turn now to a vision of the future that many are trying to implement (even if they do not share an identical vision): a global circular economy.

A World Without Waste?
The Global Circular Economy

Is it possible to have a world without waste? Where everything that is discarded can be recycled, reprocessed, or otherwise turned back into productive use, where waste is designed out of products at the outset, and where the culture of disposability has been replaced by one of thrift and reuse? This is a logical place to arrive in a book on wastes as resources and is the ultimate vision of advocates of the circular economy (see also chapter 2).[1]

I cannot imagine a world without dirt, disorder, and very bad smells. I fell in love with New York City one summer in the late 1980s, the first time I visited, to the smell of its garbage. I only learned while writing this book that New York City is the only major city in the US to pile its garbage bags up by the curb like that, a practice that is due to change. But that was a couple of decades ago, predating the rise of mass distribution of plastic bottles, lids, straws, electronic devices, accompaniments of life in the twenty-first century that are highly visible to the consumer.

There are signs that the global waste economy is internalizing elements of circularity. Incorporating smart design into products and better technology into recycling and reprocessing is a critical step. Restrictions on single-use plastic products have multiplied around the world, and many local jurisdictions are looking to integrate informal waste workers into their waste management systems (many, however, are not).

The EU's circular economy strategy stresses first and foremost its strategy for restricting plastics use only to recyclable plastics and minimizing single-use plastics. China too has developed a circular economy policy that has guided its recent 5 Year plans (Mathews and Tan 2016). In *The New Plastics Economy*, the Ellen MacArthur Foundation focused on moving away from the linear "take, make, dispose" plastics lifecycle. The NGO alliance #breakfreefromplastics brings together many organizations opposing the use and consumption of plastics altogether. Finally, many corporations, local jurisdictions, and other entities are using the circular economy framework to build a vision for their future that incorporates waste, energy, and materials use into a regenerative model, taking measures that address plastic, electronic and biological wastes.

Seeing the waste economy as a global resource frontier, with its own markets, networks, and dynamics highlights the need for a global circular economy, but also some of the problems. While the circular economy may present as a technocratic project (Gregson et al. 2015), the cases I studied in this book revealed some of the complexities and unintended negative impacts, the "shadows," cast by circular economy policies.

First, throughout the research I did for this book, one of the early tensions that emerged in terms of competing visions of a circular economy was between waste diversion and waste prevention. Should the goal be to design waste out of economic and social systems entirely (to the extent possible) or to enable and facilitate recycling, reprocessing, and energy recovery? This conflict is most visible in interactions between the waste and resource recovery industries and Circular Economy and Zero Waste advocates.

Second, there are ingrained inequities and injustices associated not only with waste generation and disposal, but with the solutions to those problems too. Enclosing landfills or establishing waste-to-energy facilities, done in the name of combating climate change or building a circular economy, can drive thousands from their livelihoods. The lack of inclusion of Southern voices in broad debates over what a global circular economy will actually mean in different parts of the world has been noted, amid concerns that as a vision it is being driven by Northern elites (Schröder et al. 2019).

Third, at what scale can circular economies be designed? Food is a case where a series of local circular economies could be designed and implemented. Food waste can be circulated and dealt with (through reduction, redistribution, composting) in the context of emerging and existing local food systems. Although these would not solve the wider structural problems that lead to food loss and waste, they could make significant progress at local levels, and do something to address the climate change implications of greenhouse gas emissions from organic waste. On the other hand, food supply chains are global, making it harder to address food loss and waste all along the chain.

In other cases, such as plastics and electronics, local solutions are unlikely to work. Just as they are manufactured in parts of the world far from where they will be used, so are recycling and reprocessing plants similarly distanced. A global or even a regional circular economy will still rely on cross-border flows of trash and recyclables, although the rather neutral term "flows" is not entirely accurate. These are decisions taken by economically motivated actors to ship discarded goods across national borders to reach markets and manufacturing centers. Yet these ongoing transactions will need some kind of governance in order to minimize the risks they pose.

These critiques come from the perspective of a political scientist steeped in the field of global governance and politics. They are intended to place these debates in a complex political context, full of competing interests, complexities and uncertainties, and justice implications, all aspects that need to be taken into account when looking to regulate this world.

However – and finally – no matter how many straws consumers in the US and Europe give up, we will never be able to reach "peak waste" (Hoornweg et al. 2013) without dealing with waste generation and disposal in the world's booming mega-cities. While waste production and landfill is reducing in Europe and the wealthier nations of the

world, non-organic waste production is increasing expo-
nentially in rapidly urbanizing parts of the global South.
So, in addition to the complexity of wastes as resources,
commodities, livelihoods, and so on, in the end we are
also coming up against a hard, age-old problem of building
good waste management systems in the cities of the world.

Final Conclusions

Wastes are universal and deeply personal. Throughout
this book are examples of how wastes, and the study of
wastes, provide lenses into broader processes and politics.
Such examples include relationships between society and
governments, from corporate sector change and growth to
street politics and civil rights activism. Wastes are under-
studied as a lens onto broader social and political issues
and conflict. The field of Discard Studies is built in part
with this mission, to unpack the significance of dirt and
disorder and situate wastes within broader social worlds
(Liboiron 2018).

As I stated in the preface, this book is not centered on
what individuals (especially concerned consumers in
Northern countries) can do about their own waste and
recycling choices and behaviors. However, and by way of
conclusion, individuals can take responsibility as citizens,
just as much, if not more so, than as consumers (and waste
producers). This book suggests several options for doing
this. One is to check what our local governments are doing
and how they need prodding or support in building adequate local waste management programs, especially in the
light of global shifts. Another is to lobby state governments
against preemptive measures that forbid local restrictions
on single-use plastics. Yet another is to support right to
repair organizations and vote, if that is one of the options,

to allow for electronic devices to be repaired. At the global level, one option is supporting waste picker associations and their goals, and understanding the political impacts as well as the environmental ones of the distancing of trash. Or find a way to work and support waste management measures at the global level. This not only helps address this particular crisis, but it is becoming apparent that waste management, recycling, and prevention policies are significant tools in the fight against climate change (O'Neill 2018b).

These ideas are far from exhaustive. In the end, there are many ways to address the global waste crisis. Understanding the opportunities, the risks, and the governance challenges of this global resource frontier provides critical context for the global community to move forward.

Notes

1 THE GLOBAL POLITICAL ECONOMY OF WASTE

1 Michael Stothard, "Nuclear Waste: Keep Out for 100,000 Years," *Financial Times*, July 14, 2016.

2 The changing relationship between the wealthy countries of the global North and the poorer countries of the global South is central to understanding the new global waste and recycling economy, even as these countries become increasingly differentiated. There are many terms for this distinction: North–South, rich–poor, developed and developing (or less developed). In international agreements, these distinctions are often formalized. In the Basel Convention on the hazardous waste trade, the treaty distinguishes between members of the Organization for Economic Cooperation and Development (OECD), representing the global North and non-OECD members representing the global South. These distinctions are changing and in practice this is never a true binary distinction. China, only 30 years ago part of the global South in global politics and an emerging economy, is now an industrialized economic superpower, as we see in chapter 6 (even though it sometimes plays the "poor country" card). Income inequality is growing within countries, a factor driving the emergence of informal work sectors in many wealthy countries. Conversely, many of the countries that are leading importers of electronic wastes, such as Ghana, are far from poor. Indeed, they have thriving industrial and consumer sectors, hence their demand for electronics wealthy Northern consumers have no use for.

2 UNDERSTANDING WASTES

1 Chapter 2 of *What a Waste 2.0* (Kaza et al. 2018; available as a free download) has useful tables and charts that depict this data. See in particular figure 2.8 on p. 27, on projected waste generation by country income group, between 2016 and 2050.

2 "Where there's muck, there's brass" is an old Yorkshire saying. Brass is money, muck is variously defined as dirt, manure, waste, or dirty business. So, where there's waste, there's money to be made. See also Paterson (2012).

3 Eurostat, Municipal Waste Statistics, Figure 2: Municipal waste treatment by type of treatment, EU-28, (kg per capita), 1995–2017 (updated January 2019) at https://ec.europa.eu/ eurostat/statistics-explained/index.php/Municipal_waste_ statistics#Municipal_waste_treatment

4 To learn more about recycling, see "A brief history of household recycling," an interactive timeline at https://www.citylab.com/ city-makers-connections/recycling/

5 For an introduction to WTE technologies, see this guide from Sustainable Energy Africa: http://sustainable.org.za/userfiles/ incineration(1).pdf

6 For more on problems with incineration and WTE, see GAIA's publications and resources at http://www.no-burn.org/ incineration/. On anti-incineration conflicts around the world, see Balkan (2012) (China); Baptista (2018) and Davies (2007) (Ireland); Rootes (2013) (UK); Henry (2018) (Moscow); Demaria and Schindler (2016) (Delhi); Sánchez (2018) (Mexico City); and many others.

7 Another term, *closed loop*, is associated with the work of industrial ecologists like Robert Frosch (e.g. 1997), referring to materials and energy flows of industrial systems (likening them to ecosystems), and is embodied in policies of extended producer responsibility (see chapter 4). It can be applied at different levels: manufacturing plant, firm, industrial sector or across industrial activities. It is most closely associated with technologies for businesses, manufacturers, and recyclers.

3 WASTE WORK

1 Dias (2016), quoting Jo Beall and Nanzeen Kanji, "Households, Livelihood and Urban Poverty," Background Paper for ESCOR commissioned research on urban development: Urban governance, partnership and poverty, 1999.
2 See US BLS statistics on Waste Management and Remediation Services (NAICS 562) at https://www.bls.gov/iag/tgs/iag562.htm#earnings and https://www.bls.gov/cps/cpsaat18.htm. It has been said that openings for New York City sanitation workers have brought in more applicants per place than Harvard University – see Jaime Hellman, "Sanitation Gold: NYC Garbage Collection Jobs in Huge Demand." *Aljazeera America*, January 13, 2015, at http://america.aljazeera.com/watch/shows/real-money-with-alivelshi/articles/2015/1/13/sanitation-gold.html
3 Data from Statista, Projected global waste management market size from 2010 to 2020 (in billion US dollars). Original file with author.
4 For further reading on informal labor in the global economy, see Ashiagbor (2019), Bob-Milliar and Obeng-Odoom (2011), Bromley and Wilson (2018), Chen (2012) and Herod (2018).
5 General references on informal waste work include Dauvergne and LeBaron (2013), Dias (2016), Dias and Samson (2012), Grant and Oteng-Ababio (2012, 2016), Hartmann (2018), Linzner and Lange (2012), Marello and Helwege (2018), Moore (2009), Parizeau (2015), and Vergara and Tchobanoglous (2012).
6 On the Zabaleen in Cairo, see Bower (2017); Davies (2016); Fahmi and Sutton (2010); Hessler (2014); Slackman (2009).
7 See Bureau of Labor Statistics, and the NYC Department of Sanitation's 2016 Private Carting Study, reported in Rosengren (2016).

4 DISCARDED ELECTRONICS

1 J.B. MacKinnon, "The L.E.D. Quandary: Why there's no such thing as 'built to last'," *The New Yorker*, July 14, 2016. See also G. Slade, *Made to Break: Technology and Obsolescence in America.* Cambridge, MA: Harvard University Press, 2009.

2 See the USEPA's e-waste recycling data at https://www.epa.gov/
recycle/electronics-donation-and-recycling.

3 For more on urban mining, see Grant and Oteng-Ababio (2012,
2016), Knapp (2016), Krook and Baas (2013), Lepawsky et al.
(2017), Minter (2013); Labban (2014) and Reddy (2016).

4 This section dramatically oversimplifies the different grades
and types of scrap. For the distinctions between "new" and
"old" scrap, and such industry specific terms as zorba, zurik,
and zebra, see ISRI's Scrap Specifications Circular, available on
ISRI's website, Minter (2013), and Graedel et al. (2011).

5 HS Codes, "Harmonized commodity description and coding
system" codes, standardize what is being traded throughout
import and export records. Trade data can be downloaded at the
UN Comtrade database at https://comtrade.un.org.

6 Adam Minter (@adamminter), "Lots of News These Days
About the Current Global Recycling Crisis," Twitter,
September 13, 2018, https://twitter.com/AdamMinter/
status/1040178845381324800

5 FOOD WASTE

1 Established in 1945, the UNFAO's mandate is to end global
hunger. It is also the leading source of expertise and authoritative
information on all aspects of food and agriculture worldwide.

2 For more on WRAP's mission and work, see http://www.wrap.
org.uk. WRAP also works on plastic wastes and the circular
economy, doing research and advocacy connecting these
problems and providing policy recommendations.

3 On the history of food waste, see Schneider (2013) and Evans et
al. (2013).

4 See Eric Holthaus, "Stop Buying in Bulk," at https://slate.com/
business/2015/06/bulk-shopping-creates-food-waste-shop-
more-often-instead.html

5 Food waste activist organizations are strongly associated with
food justice, security and sovereignty activism, and the growth
of local/alternative food movements; see Alkon and Guthman
(2017). For groups on the ground, see "59 Organizations
Fighting Food Loss and Waste," at https://foodtank.com/
news/2016/07/fighting-food-loss-and-waste/

6 ReFED's map of innovators worldwide is at https://www. refed.com/tools/innovator-database/innovators#active_ tab=innovatorMap

7 See Gunders (2012), Gunders et al. (2017), and The World Resource Institute's Food Loss and Waste Protocol and related pages at http://www.flwprotocol.org

8 Laudan's blog post can be found at https://www.rachellaudan. com/2017/03/im-a-happy-food-waster.html. See also Alexander et al. (2013) and Gjerris and Gaiani (2013), and for a satirical perspective from *The Onion* on how to cut down food waste, see https://www.theonion.com/how-to-cut-down-on-food-waste-1819592516.

9 Claire Provost and Felicity Lawrence, "US food aid program criticized as 'corporate welfare' for grain giants." *The Guardian*, July 18, 2012; Mark Tran, "EU agriculture policy 'still hurting farmers in developing countries'." *The Guardian*, October 11, 2011.

6 PLASTIC SCRAP

1 These are just a few of the many books and articles on plastics. Writer Rebecca Altman maintains a list on her website, http:// rebecca-altman.com and see also many others of the works cited in this chapter.

2 https://www.epa.gov/facts-and-figures-about-materials-waste-and-recycling/plastics-material-specific-data

3 This figure was reported in Recycling Today on July 7, 2014, reporting data from the China Scrap Plastics Association (see http://www.recyclingtoday.com/article/china-plastic-recycling-rate-decline/).

4 According to ISRI's brief filed to the WTO and China's Ministry of Environmental Protection, Notification G/TBT/N/CHN/1211, August 2017.

5 Plastic and paper scrap (HS 3915 and HS 4707) were the most high-profile of the 24 types, given their impact on the post-consumer recycling industry. The other types include slag, dross, and other waste from manufacture of iron or steel (HS 2619), ash and residues containing arsenic, metals or their compounds (HS 2620); waste of wool or of animal hair,

including yarn waste (HS 5103); cotton waste (HS 5202); waste of artificial fibers (HS 5505); used, new and worn-out rags, scrap twine, rope, cables (HS 6310).

6 Xavier A. Cronin, "America's Plastic Scrap Draft," *Recycling Today*, September 30, 2016, at http://www.recyclingtoday.com/ article/americas--plastic-scrap-draft/. Little is known about exports to the Heard and the McDonald Islands cited in this piece.

7 *Waste Dive*, "How recycling is changing in all 50 states," at https://www.wastedive.com/news/what-chinese-import-policies-mean-for-all-50-states/510751/

8 These lists also include cutlery, cotton buds, and food containers. On bottles, see Hawkins et al. (2015).

9 For an account of the history and present of the plastic bag, see Rebecca Altman, "American Beauties." *Topic*, August 2018, at https://www.topic.com/american-beauties

10 For up-to-date information, see http://www.ncsl.org/research/ environment-and-natural-resources/plastic-bag-legislation.aspx

11 On plastic straws, see Sarah Gibbens, "A brief history of how plastic straws took over the world," *National Geographic* "Planet or Plastics" issue, June 2018, and for a contrary opinion, Minter (2018).

12 Biodegradable and compostable are two separate attributes. Everything is biodegradable in the end, even plastics, we think. "Compostable" is the better term to look for on packaging, etc. This means that under the right conditions it will break down entirely. However, sometimes the conditions required are quite specialized: some compostable cutlery, for example, may take longer to decompose than the facility they are sent to can handle.

13 For resources on this topic, see https://www.ciel.org/plastic-waste-proposal-basel-convention/

14 The WTO incorporated the GATT and other related agreements in 1995, to create a stronger organization more able to manage the complexity of global trade in goods and services.

15 One of the best sources on WTO disputes is the WTO website itself. It contains detailed descriptions of each case, arguments and outcomes. See https://www.wto.org/english/ tratop_e/dispu_e/dispu_e.htm. The Center for International Environmental Law (CIEL) also tracks these cases.

16 See the Center for International Environmental Law's analysis at https://www.ciel.org/project-update/brazil-retreaded-tires/

CONCLUSION: A WORLD WITHOUT WASTE?

1 James Quincey, the President and CEO of The Coca Cola Company gave a speech on this very topic at the 2018 World Economic Forum: "Moving toward a world without waste," available at https://www.weforum.org/agenda/2018/01/world-without-waste-recycle-plastic/

Selected Readings:
Making Wastes Visible

Finding out more about wastes is so much easier than it was 20 years ago when I first worked on this topic. Reports, news services, academic literature, and general non-fiction are all far more plentiful. Recent studies, activist reports, journalism, and scholarship have served to make wastes more visible, and wastes' journeys more traceable. Daily bulletins such as *Waste Dive*, reports by major trade associations and international organizations, and work in the field of discard studies have made waste data, news, and scholarly analyses more plentiful and accessible than when I started in this field. Many of the organizations, journals, and writers mentioned in this section and in this book are active on Twitter, one of the best ways to keep up with the field (e.g. @discardstudies, @wastedive, @reassemblingrubbish, and many more).

This selected readings section focuses on general reading and directions for finding out more about waste and doing research in this field. But first, while basic waste data is easier to find now, caveats apply. Many of the works listed below discuss limitations with their data: different definitions, incomplete information, withheld information, and other issues challenge transparency in waste production, disposal, and governance. Still, all the works cited in this section have contributed to making wastes visible – as a risk, a resource, a livelihood, and much more. All readings are listed in full in the References section of this book.

Influential reports on the global waste situation now and into the future include *What a Waste* (2012 and 2018), published by the World Bank. It introduced the concept of "peak waste" and explores waste trends across countries and over time. The *Global Waste Management Outlook* (2015), a 300-page report from the UN Environment Programme and the International Solid Waste Association remains a go-to source. Likewise, the *Global E-Waste Monitors* (Baldé et al. 2015 and 2017) provide snapshots of electronic waste production around the world. The NRDC and the UNFAO have produced the most comprehensive reports on food waste and loss (e.g. Gunders 2012; UNFAO 2011). On plastics, the work of Jenna Jambeck at the University of Georgia and her collaborators are setting the global standard (Jambeck et al. 2015; Geyer et al. 2017; Brooks et al. 2018).

Waste reporting is thriving. *Waste Dive* is published daily (www.wastedive.com), and covers all the important stories, from global events, such as its ongoing coverage of China's Operation National Sword, to the local, even the personal (e.g. studies that show women recycle more than men). *Waste 360*, *Recycling Today*, and *Waste Management World* are trade journals that also provide news reporting services; a good way to find out about the industry. For general stories, *The Guardian* newspaper provides excellent coverage of waste, food waste, and plastics-related issues.

Waste as a topic lends itself particularly well to writing for general audiences. Influential works include Robin Nagle's *Picking Up: On the Streets and Behind the Trucks with the Sanitation Workers of New York City* (2013), Adam Minter's *Junkyard Planet: Travels in the Billion Dollar Trash Trade* (2013), Carl Zimring's *Clean and White: A History of Environmental Racism in the United States* (2015), Samantha MacBride's *Recycling Reconsidered* (2012), Joshua Reno's *Waste Away: Working and Living with a North American*

Landfill (2016), and Susan Strasser's *Waste and Want: A Social History of Trash* (1999). On Garbology and the importance of trash in understanding our history, see Humes (2010) and Rathje and Murphy (2001). Other books or articles that are "must reads" include, as just a sampling, Lepawsky's *Reassembling Rubbish* (2018), David Wilson's 2007 article on the history of waste management, and Andrew Szasz's 1986 article on hazardous waste and organized crime. Classics from the late 1980s and early 1990s on the global political economy of waste include Gourlay's *World of Waste* (1992), the work of Brian Wynne (e.g. 1987 and 1989), and Mary Douglas's *Purity and Danger* (1966), a foundational text for the Discard Studies field. Of the earliest works on my list, I highlight Atwater (1895), a very early account of food waste.

The Discard Studies website (www.discardstudies.com) is the best curated collection of work on "wider systems, structures and cultures of waste and wasting," covering art, politics, the humanities, social sciences, engineering, and many other perspectives on this topic. The site's founder, Max Liboiron, has written extensively on Discard Studies as a field (e.g. Liboiron 2018, "The What and the Why of Discard Studies"), as well as an article that demonstrates how the concept of "discard" helps us understand other social phenomena ("Tactics of Waste, Dirt and Discard in the Occupy Movement," 2012). In this book, I was not able to cover these broader applications of "discards," but see also Gidwani and Reddy, "The Afterlives of 'Wastes'" (2011).

Collecting qualitative data, especially on informal waste work, poses logistical and ethical challenges. Raul Pacheco-Vega and Kate Parizeau have written on how to work with vulnerable communities using their own work on waste picker communities as examples, deploying the concept of "doubly-engaged ethnography" (Pacheco-Vega and Parizeau 2018).

Multimedia presentations and sources on waste are another way in which wastes are becoming more visible. Annie Leonard's *The Story of Stuff* is an online animated documentary that has mobilized audiences worldwide, and led to an ongoing multimedia project. On urban mining and Ghana, DK Osseo-Asare, a professor and practitioner of architecture and design recorded a TED talk on Agbogbloshie, Ghana in 2017, "What a scrapyard in Ghana can teach us about innovation" which can be found on YouTube at https://www.youtube.com/watch?v=i_wtaoHCw3k. Tammara Soma has a video on YouTube on food waste in Indonesia, at https://www.youtube.com/watch?v=9q46XBJNbP4. Suez Environnement has on its website a short video that untangles what made today's complex waste management corporations and has original footage of early wastewater and solid waste collection systems going back to the nineteenth century, at https://www.suez.com/en/Who-we-are/A-worldwide-leader/Our-history. "A brief history of household recycling," is an updated interactive timeline at https://www.citylab.com/city-makers-connections/recycling/.

This book situated the world of Waste and Discard Studies in other broader fields, notably Global Environmental Governance and Global Political Economy. For a reader interested in finding out more about these fields, there are several places to start. First, the chapters in *Confronting Consumption* (Princen et al. 2002) help us understand the concept of distancing, and the neglected role of consumption in global politics and governance. Peter Dauvergne's 2018 article on "How Governance of Plastics is Failing the Ocean" outlines how traditional tools of global environmental governance could be applied in the plastics case. O'Neill's *The Environment and International Relations* (2017b) is a thorough introduction to the basics of global environmental politics and governance. Private global governance came up several times, and I would recommend the work of Benjamin Cashore and colleagues on the ins and outs of certification systems (e.g. Cashore et al. 2004). For additional work on resource frontiers, see Peluso (2017). On the changing role of China in the world, see Zinda et al. (2018).

Finally, no section that deals with making wastes visible could leave out wastes and waste work and their representation in popular culture. Arts and literature explore human relationships with wastes through visualization and fiction. Waste art – ensembles of old circuit boards, sculptures made from ocean plastics, photos of dumps made to look like classic Chinese art – is practically its own genre and widely available. Waste-related themes abound in science fiction and other genres (*Shipbreaker* by Paolo Bacigalupi, and the *Great Pacific* series of graphic novels by Joe Harris and Martin Morazzo). Scavengers and scrap traders are a popular choice of "outsider" in movie depictions (*Mad Max: Fury Road*, *Firefly*, and of course *Star Wars*). And last but not least, Pixar's *Wall-E* (2008), a favorite movie for many, starts in a huge junkyard, the depleted Earth humans had left.

References

Acaroglu, Leyla. "System Failures: Planned Obsolescence and Enforced Disposability." *Medium.com*, May 7, 2018.

Alexander, Catherine, Nicky Gregson, and Zsuzsa Gille. "Food Waste." In *The Handbook of Food Research*. Eds. Murcott, Anne, Warren Bellasco and Peter Jackson. London: Bloomsbury, 2013, pp. 471–84.

Alkon, Alison, and Julie Guthman, eds. *The New Food Activism: Opposition, Cooperation, and Collective Action*. Berkeley, CA: University of California Press, 2017.

Altman, Rebecca. "American Petro-Topia – An Intimate History of Plastic." *Aeon Magazine*, March 2015.

Arora, Rachna, Katharina Paterok, Abhijit Banerjee, and Manjeet Singh Saluja. "Potential and Relevance of Urban Mining in the Context of Sustainable Cities." *IIMB Management Review* 29 (2017): 210–24.

Aschemann-Witzel, Jessica. "Waste Not, Want Not, Emit Less." *Science* 352.6284 (2016): 408–9.

Aschemann-Witzel, Jessica, Ilona de Hooge, and Anne Normann. "Consumer-Related Food Waste: Role of Food Marketing and Retailers and Potential for Action." *Journal of International Food & Agribusiness Marketing* 28.3 (2016): 271–85.

Ashiagbor, Diamond, ed. *Re-Imagining Labor Law for Development: Informal Work in the Global North and South*. London: Hart Publishing, 2019.

Atwater, W. O. *Methods and Results on Investigations on the Chemistry and Economy of Food*. Bulletin No. 21, US Department of Agriculture, Office of Experiment Stations. Washington, DC: Government Printing Office, 1895.

Australian Packaging Covenant Organisation. *Market Impact Assessment Report: Chinese Import Restrictions for Packaging in Australia*. March 30, 2018, available at https://www.packagingcov enant.org.au/documents/item/1224

Baldé, C.F., V. Forti, V. Gray, R. Kuehr, and P. Stegmann. *The Global E-Waste Monitor – 2017*. Bonn/Geneva/Vienna: United Nations University (UNU), International Telecommunication Union (ITU) and International Solid Waste Association (ISWA), 2017.

Baldé, C.F., F. Wang, R. Kuehr, and J. Huisman. *The Global E-Waste Monitor – 2014*. Bonn: United Nations University; IAS, 2015.

Balkan, Elizabeth. "Dirty Truth About China's Incinerators." *China Dialogue*, May 7, 2012.

Baptista, Ana. "Garbage in, Garbage Out: Incinerating Trash Is Not an Effective Way to Protect the Climate or Reduce Waste." *The Conversation*, February 27, 2018.

Barbier, Edward B. "Scarcity, Frontiers and the Resource Curse: A Historical Perspective." In *Natural Resources and Economic Growth: Learning from History*. Eds. Badia-Miró, Marc, Vicente Pinilla and Henry Willebald. London: Routledge, 2015, pp. 54–76.

Barboza, Luís Gabriel Antão, A. Dick Vethaak, Beatriz R.B.O. Lavorante, Anne-Katrine Lundebyef, and Lúcia Guilhermino. "Marine Microplastic Debris: An Emerging Issue for Food Security, Food Safety and Human Health." *Marine Pollution Bulletin* 133 (2018): 336–48.

Barnard, Alex V. "'Waving the Banana' at Capitalism: Political Theater and Social Movement Strategy among New York's 'Freegan' Dumpster Divers." *Ethnography* 12.4 (2011): 419–44.

Basel Action Network. *Scam Recycling: E-Dumping on Asia by US Recyclers*. Seattle, WA: Basel Action Network e-Trash Transparency Project, 2016.

Beall, Jo. "Thoughts on Poverty from a South Asian Waste Dump: Gender, Inequality and Household Waste." *IDS Bulletin* 28.3 (1997): 73–90.

Bhatia, Juhie. "Ugly Fruits and Vegetables: Why You Have to Learn to Love Them." *The Guardian*, November 17, 2016.

BIPRO/CRI. *Assessment of Separate Collection Schemes in the 28 Capitals of the EU, Final Report*. Brussels: European Commission – DG ENV, 2015.

Bloch, Stefano. "Hollywood as Waste Regime." *City* 17.4 (2013): 449–73.

Block, Alan A., and Frank R. Scarpitti. *Poisoning for Profit: The Mafia and Toxic Waste in America*. New York: William Morrow & Co, 1985.

Bob-Milliar, George M., and Franklin Obeng-Odoom. "The Informal Economy Is an Employer, a Nuisance, and a Goldmine: Multiple

Representations of and Responses to Informality in Accra, Ghana." *Urban Anthropology and Studies of Cultural Systems and World Economic Development* 40.3/4 (2011): 263–84.

Boo, Katherine. *Behind the Beautiful Forevers: Life, Death, and Hope in a Mumbai Undercity*. New York: Random House, 2012.

Boteler, Cody. "Interpol Seizes More Than 1.5m Metric Tons of Illegal Waste in 30-Day Operation." *Waste Dive*, August 15, 2017.

Bower, Edmund. "One Man's Trash," *Business Monthly*, May 2017.

Breivik, Knut, James M. Armitage, Frank Wania, and Kevin C. Jones. "Tracking the Global Generation and Exports of E-Waste: Do Existing Estimates Add Up?" *Environmental Science and Technology* 48 (2014): 8735–43.

Bromley, Ray, and Tamar Diana Wilson. "Introduction: The Urban Informal Economy Revisited." *Latin American Perspectives* 45.1 (2018): 4–23.

Brooks, Amy L., Shunli Wang, and Jenna R. Jambeck. "The Chinese Import Ban and Its Impact on Global Plastic Waste Trade." *Science Advances* 4 (2018): 1–7.

Brooks, Andrew. "Networks of Power and Corruption: The Trade of Japanese Used Cars to Mozambique." *The Geographical Journal* 178.1 (2012): 80–92.

Brooks, Andrew. "Stretching Global Production Networks: The International Second-Hand Clothing Trade." *Geoforum* 44 (2013): 10–22.

Bullard, Robert. *Dumping in Dixie: Race, Class and Environmental Quality*. Boulder, CO: Westview Press, 1991.

Carlin, Aine. "We All Love Bagged Salads but They're the Tip of the Food Waste Iceberg." *The Guardian*, May 24, 2017.

Cashore, Benjamin, Graeme Auld, and Deanna Newsom. *Governing through Markets: Forest Certification and the Emergence of Non-State Authority*. New Haven, CT: Yale University Press, 2004.

Castaldi, Marco J. "Perspectives on Sustainable Waste Management." *Annual Review of Chemical and Biomolecular Engineering* 5 (2014): 547–62.

Chambers, Robert, and Gordon R. Conway. "Sustainable Rural Livelihoods: Practical Concepts for the 21st Century." IDS Discussion Paper 296 (1991).

Chen, Martha Alter. "The Informal Economy: Definitions, Theories and Policies." WIEGO Working Papers (2012).

Ciplet, David. "Contesting Climate Injustice: Transnational Advocacy Network Struggles for Rights in UN Climate Politics." *Global Environmental Politics* 14.4 (2014): 75–96.

Clapp, Jennifer. "The Distancing of Waste: Overconsumption in a Global Economy." In *Confronting Consumption*. Eds. Princen, Thomas, Michael F. Maniates and Ken Conca. Cambridge, MA: MIT Press, 2002.

Clapp, Jennifer. *Hunger in the Balance: The New Politics of International Food Aid.* Ithaca, NY: Cornell University Press, 2012.

Clapp, Jennifer. *Food* (2nd edn). London: Polity, 2016.

Clapp, Jennifer, and Linda Swanston. "Doing Away with Plastic Shopping Bags: International Patterns of Norm Emergence and Policy Implementation." *Environmental Politics* 18.3 (2009): 315–32.

Coffin, David, Jeff Horowitz, Danielle Nesmith, and Mitchell Semanik. "Examining Barriers to Trade in Used Vehicles." Office of Industries Working Paper, US International Trade Commission ID-044 (2016).

Crooks, Harold. *Giants of Garbage: The Rise of the Global Waste Industry and the Politics of Pollution Control.* Toronto: James Lorimer and Company, 1993.

Dauvergne, Peter. "Why Is the Global Governance of Plastic Failing the Oceans?" *Global Environmental Change* 61 (2018): 22–31.

Dauvergne, Peter, and Genevieve LeBaron. "The Social Cost of Environmental Solutions." *New Political Economy* 18.3 (2013): 410–30.

Davies, Anna R. "Civil Society Activism and Waste Management in Ireland: The Carranstown Anti-Incineration Campaign." *Land Use Policy* 25 (2007): 161–72.

Davies, Caroline. "Garbage City: The Scavengers Making a Fortune from Other Peoples' Rubbish," *BBC News Magazine*, September 26, 2016.

Davis, Shannon. "The Commodification of Waste: A Story of Plastics Recycling in Berkeley." Honors Thesis, Society and Environment Major. University of California at Berkeley, 2014.

Demaria, Federico, and Seth Schindler. "Contesting Urban Metabolism: Struggles over Waste-to-Energy in Delhi, India." *Antipode* 48.2 (2016): 293–313.

Dias, Sonia Maria. "Waste Pickers and Cities." *Environment & Urbanization* 28.2 (2016): 375–90.

Dias, Sonia Maria. "4 Strategies to Integrate Waste Pickers into Future Cities." *WIEGO Blog*, February 28, 2017.

Dias, Sonia Maria. "Statistics on Waste Pickers in Brazil." WIEGO Statistical Brief, No. 2 (2011).

Dias, Sonia Maria, and Melanie Samson. *Informal Economy Monitoring Sector Report: Waste Pickers*. Cambridge, MA: WIEGO, 2012.

Dinler, Demet S. "New Forms of Wage Labor and Struggle in the Informal Sector: The Case of Waste Pickers in Turkey." *Third World Quarterly* 37.10 (2016): 1834–54.

Douglas, Mary. *Purity and Danger*. London: Routledge and Kegan Paul, 1966.

Earley, Katharine. "How Technology Can Prevent Food Waste in Developing Countries." *The Guardian*, December 18, 2014.

Economist. "A 'right to repair' movement tools up." *The Economist*, September 30, 2017.

Ellen MacArthur Foundation. *A New Textiles Economy: Redesigning Fashion's Future*. 2017.

Elliott, Lorraine. "Governing the International Political Economy of Transnational Environmental Crime." In *Handbook of the International Political Economy of Governance*. Eds. Payne, Anthony and Nicola Phillips. Cheltenham: Edward Elgar Publishing, 2014, pp. 450–68.

European Environment Agency. "Environmental Goods and Services Sector: Employment and Value Added". EEA, 2018. Available at https://www.eea.europa.eu/airs/2018/resource-efficiency-and-low-carbon-economy/environmental-goods-and-services-sector

Evans, David, Hugh Campbell, and Anne Murcott. "A Brief Pre-History of Food Waste and the Social Sciences." *The Sociological Review* 60.S2 (2013): 5–26.

Evans, David, Hugh Campbell, and Anne Murcott, eds. *Waste Matters: New Perspectives on Food and Society*. Hoboken, NJ: Wiley, 2013.

Fahmi, Wael, and Keith Sutton. "Cairo's Contested Garbage: Sustainable Solid Waste Management and the Zabaleen's Right to the City." *Sustainability* 2 (2010): 1765–83.

Flower, Will. "What Operation Green Fence Has Meant for Recycling." *Waste360*, February 10, 2016.

Freinkel, Susan. *Plastics: A Toxic Love Story*. New York: Houghton Mifflin Harcourt, 2011.

Friedman, Gerard. "Workers without Employers: Shadow Corporations and the Rise of the Gig Economy." *Review of Keynesian Economics* 2.2 (2014): 171–88.

Frosch, Robert A. "Industrial Ecology: Closing the Loop on Waste Materials." In *The Industrial Green Game*. Ed. Richards, Deanna J. Washington, DC: National Academies Press, 1997.

Gardner, T. A., M. Benzie, J. Börner, et al. "Transparency and Sustainability in Global Commodity Supply Chains." *World Development* (2018). https://doi.org/10.1016/j.worlddev.2018.05.025

Geeraerts, Kristof, Andrea Illes, and Jean-Pierre Schweizer. *Illegal Shipment of E-Waste from the EU: A Case Study on Illegal Export from the EU to China*. London: IEEP, 2015.

Geoghegan, Tom. "The Story of How the Tin Can Nearly Wasn't." *BBC News Magazine*, April 21, 2013.

Geyer, Roland, Jenna R. Jambeck, and Kara Lavender Law. "Production, Use, and Fate of All Plastics Ever Made." *Science Advances* 3.7 (2017): 1–5.

Gibbs, Carole, Edmund F. McGarrell, Mark Axelrod, and Louie Rivers III. "Conservation Criminology and the Global Trade in E-Waste: Applying a Multi-Disciplinary Research Framework." *International Journal of Comparative and Applied Criminal Justice* 35.4 (2011): 269–91.

Gidwani, Vinay. "The Work of Waste: Inside India's Infra-Economy." *Transactions of the Institute of British Geographers* 40.4 (2015): 575–95.

Gidwani, Vinay, and Rajyashree N. Reddy. "The Afterlives of 'Wastes': Notes from India for a Minor History of Capitalist Surplus." *Antipode* 43.5 (2011): 1625–58.

Gille, Zsuzsa. "From Risk to Waste: Global Food Waste Regimes." *The Sociological Review* 60.S2 (2012): 27–46.

Gjerris, Mickey, and Silvia Gaiani. "Household Food Waste in Nordic Countries: Estimations and Ethical Implications." *Nordic Journal of Applied Ethics* 7.1 (2013): 6–23.

Gourlay, K.A. *World of Waste: Dilemmas of Industrial Development*. London: Zed Books, 1992.

Graedel, Thomas E., Julian Allwood, Jean-Pierre Birat, et al. "What Do We Know About Metal Recycling Rates?" *Journal of Industrial Ecology* 15.3 (2011): 355–66.

Grant, Kristen, Fiona C. Goldizen, Peter D. Sly, et al. "Health

Consequences of Exposure to E-Waste: A Systematic Review." *Lancet Global Health* 1 (2013): 350–61.

Grant, Richard, and Martin Oteng-Ababio. "Mapping the Invisible and Real 'African' Economy: Urban E-Waste Circuitry." *Urban Geography* 33.1 (2012): 1–21.

Grant, Richard, and Martin Oteng-Ababio. "The Global Transformation of Materials and the Emergence of Urban Mining in Accra, Ghana." *Africa Today* 62.4 (2016): 2–20.

Gregson, Nicky, Mike Crang, Julie Botticello, Melania Calestani, and Anna Krzwoszynska. "Doing the 'Dirty Work' of the Green Economy: Resource Recovery and Migrant Labor in the EU." *European Urban and Regional Studies* 23.4 (2016): 541–55.

Gregson, Nicky, Mike Crang, Sara Fuller, and Helen Holmes. "Interrogating the Circular Economy: The Moral Economy of Resource Recovery in the EU." *Economy and Society* 44.2 (2015): 218–43.

Gunders, Dana. *Wasted: How America Is Losing up to 40 Percent of Its Food from Farm to Fork to Landfill*. New York: National Resources Defense Council, 2012.

Gunders, Dana, Jonathan Bloom, JoAnne Berkenkamp, Darby Hoover, Andrea Spacht, and Marie Mourad. *Wasted: How America Is Losing up to 40 Percent of Its Food from Farm to Fork to Landfill, Second Edition of NRDC's Original 2012 Report*. New York: National Resources Defense Council, 2017.

Hartmann, Chris. "Waste Picker Livelihoods and Inclusive Neoliberal Municipal Solid Waste Management Policies: The Case of the La Chureca Garbage Dump Site in Managua, Nicaragua." *Waste Management* 71 (2018): 565–77.

Havice, Elizabeth, and Kristin Reed. "Fishing for Development? Tuna Resource Access and Industrial Change in Papua New Guinea." *Journal of Agrarian Change* 12.2–3 (2012): 413–35.

Hawkins, Gay, Emily Potter, and Kane Race. *Plastic Water: The Social and Material Life of Bottled Water*. Cambridge, MA: MIT Press, 2015.

Heneghan, Carolyn. "Why Tackling Food Waste Is a Win-Win for Manufacturers – and 3 Ways It's Getting Done." *Waste Dive*, June 27, 2016.

Henry, Laura. "Will a Garbage Revolt Threaten Putin?" *The Conversation*, June 7, 2018.

Henz, Gilmar Paulo, and Gustavo Porpino. "Food Losses and

Waste: How Brazil Is Facing This Global Challenge?" *Horticultura Brasileira* 35.4 (2017): 472–82.

Herod, Andrew. *Labor*. London: Polity, 2018.

Hessler, Peter. "Tales of the Trash", *The New Yorker*, October 13, 2014.

Hobson, Kersty, and Nicholas Lynch. "Diversifying and De-Growing the Circular Economy: Radical Social Transformation in a Resource-Scarce World." *Futures* 82 (2016): 15–25.

Hoornweg, Daniel, and Perinaz Bhada-Tata. *What a Waste: A Global Review of Solid Waste Management*. Washington, DC: World Bank, 2012.

Hoornweg, Daniel, Perinaz Bhada-Tata, and Chris Kennedy. "Waste Production Must Peak This Century." *Nature* 502 (2013): 615–17.

Hopewell, Jefferson, Robert Dvorak, and Edward Kosior. "Plastics Recycling: Challenges and Opportunities." *Philosophical Transactions of the Royal Society* 364 (2009): 2115–26.

Howard, Brian Clark. "What Do Recycling Symbols on Plastics Mean?" *Good Housekeeping*, November 25, 2008.

Humes, Edward. *Garbology: Our Dirty Love Affair with Trash*. New York: Penguin Group, 2010.

Iles, Alastair. "Mapping Environmental Justice in Technology Flows: Computer Waste Impacts in Asia." *Global Environmental Politics* 4.4 (2004): 78–107.

International Energy Agency. *The Future of Petrochemicals: Towards More Sustainable Plastics and Fertilizers*. Paris: International Energy Agency, 2018.

ISWA. *Globalization and Waste Management, Phase 1: Concepts and Facts*. Vienna: International Solid Waste Association, 2012.

ISWA. *A Roadmap for Closing the World's Largest Dumpsites: The World's Most Polluted Places*. Vienna: International Solid Waste Association, 2016.

Jambeck, Jenna R., Roland Geyer, Chris Wilcox, et al. "Plastic Waste Inputs from Land into the Ocean." *Science* 347.6223 (2015): 768–71.

Jeffries, Nick. "A Circular Economy for Food: 5 Case Studies." *Circulate News*, February 9, 2018.

Jolly, Janice L. *The US Copper-Based Scrap Industry and Its Byproducts: An Overview* (7th edn). New York: Copper Development Association, 2007.

Kagan, Robert A. *Adversarial Legalism: The American Way of Law*. Cambridge, MA: Harvard University Press, 2001.

Kaza, Silpa, Lisa Yao, Perinaz Bhada-Tata, and Frank Van Woerden. *What a Waste 2.0: A Global Snapshot of Solid Waste Management.* Washington, DC: World Bank, 2018.

Khalamayzer, Anya. "Ikea's 7 Imperatives for Scrapping Food Waste." *GreenBiz,* July 24, 2017.

Kiser, Barbara. "Getting the Circulation Going." *Nature* 531 (2016): 443–44.

Knapp, Freyja L. "The Birth of the Flexible Mine: Changing Geographies of Mining and the E-Waste Commodity Frontier." *Environment and Planning A* 48.10 (2016): 1–21.

Knoblauch, Jessica A. "Plastic Not-So-Fantastic: How the Versatile Material Harms the Environment and Human Health." *Environmental Health News,* July 2, 2009.

Krook, Joakim, and Leenard Baas. "Getting Serious About Mining the Technosphere: A Review of Recent Landfill Mining and Urban Mining Research." *Journal of Cleaner Production* 55 (2013): 1–9.

Kumar, Deepak, and Prasanta Kalita. "Reducing Post-Harvest Losses During Storage of Grain Crops to Strengthen Food Security in Developing Countries." *Food* 6.1 (2017): 8.

Labban, Mazen. "Deterritorializing Extraction: Bioaccumulation and the Planetary Mine." *Annals of the Association of American Geographers* 104.3 (2014): 560–76.

Laville, Sandra. "Chinese Ban on Plastic Waste Imports Could See UK Pollution Rise." *The Guardian,* December 7, 2017.

Lebreton, L., B. Slat, F. Ferrari, et al. "Evidence That the Great Pacific Garbage Patch Is Rapidly Accumulating Plastic." *Scientific Reports* 8.4666 (2018).

Lepawsky, Josh. *Reassembling Rubbish: Worlding Electronic Waste.* Cambridge, MA: MIT Press, 2018.

Lepawsky, Josh, Erin Araujo, John-Michael Davis, and Ramzy Kahhat. "Best of Two Worlds? Towards Ethical Electronics Repair, Reuse, Repurposing and Recycling." *Geoforum* 81 (2017): 87–99.

Lepawsky, Josh, and Chris McNabb. "Mapping International Flows of Electronic Waste." *The Canadian Geographer* 54.2 (2009): 177–95.

Levitan, Dave. "Recycling's Final Frontier: The Composting of Food Waste." *Yale 360,* August 2013.

Li, Yun, Xingang Zhao, Yanbin Li, and Xiaoyu Li. "Waste Incineration Industry and Development Policies in China." *Waste Management* 46 (2015): 234–41.

Liboiron, Max. "Tactics of Waste, Dirt and Discard in the Occupy Movement." *Social Movement Studies* 11.3–4 (2012): 393–401.

Liboiron, Max. "Municipal Versus Industrial Waste: Questioning the 3-97 Ratio." *Discard Studies*, March 2, 2016.

Liboiron, Max. "The What and the Why of Discard Studies." *Discard Studies*, September 1, 2018.

Linzner, Roland, and Ulrike Lange. "Role and Size of Informal Sector in Waste Management – A Review." *Waste and Resource Management* 166.2 (2012): 69–83.

Lundgren, Karin. *The Global Impact of E-Waste: Addressing the Challenge*. Geneva: International Labor Organization, SECTOR and SAFEWORK offices, 2012.

MacBride, Samantha. *Recycling Reconsidered: The Present Failure and Future Promise of Environmental Action in the United States*. Cambridge, MA: MIT Press, 2012.

McKee, Emily. "Trash Talk: Interpreting Morality and Disorder in Negev/Naqab Landscapes." *Current Anthropology* 56.5 (2015): 733–52.

McKenzie, Tara, Lila Singh-Peterson, and Steen J.R. Underhill. "Quantifying Postharvest Loss and the Implication of Market-Based Decisions: A Case Study of Two Commercial Domestic Tomato Supply Chains in Queensland, Australia." *Horticulturae* 3.44 (2017): 1–15.

Maes, T., J. Barry, H.A. Leslie, et al. "Below the Surface: Twenty-Five Years of Seafloor Litter Monitoring in Coastal Seas of North West Europe (1992–2017)." *Science of the Total Environment* 630 (2018): 790–8.

Marello, Marta, and Ann Helwege. "Solid Waste Management and Social Inclusion of Waste Pickers: Opportunities and Challenges." *Latin American Perspectives* 45.1 (2018): 108–29.

Matchar, Emily. "The Fight for the Right to Repair." *Smithsonian Magazine*, July 13, 2016.

Mathews, John A., and Hao Tan. "Lessons from China." *Nature* 531 (2016): 440–2.

Meikle, Jeffrey. *American Plastic: A Cultural History*. New Brunswick, NJ: Rutgers University Press, 1995.

Minter, Adam. *Junkyard Planet: Travels in the Billion-Dollar Trash Trade*. New York: Bloomsbury, 2013.

Minter, Adam. "How We Think of E-Wastes Is in Need of Repair." *Anthropocene Magazine*, October 2016.

Minter, Adam. "China's War on Foreign Garbage." *Bloomberg.com*, July 20, 2017.

Minter, Adam. "Plastic Straws Aren't the Problem." *Bloomberg Opinion*, June 7, 2018.

Mitchell, Scott. "Narratives of Resistance and Repair in Consumer Society." *Third Text* 32.1 (2018): 56–67.

Montgomery, Mark A. "Reassessing the Waste Trade Crisis: What Do We Really Know?" *Journal of Environment and Development* 4.1 (1995): 1–28.

Moore, Sarah A. "The Excess of Modernity: Garbage Politics in Oaxaca, Mexico." *The Professional Geographer* 61.4 (2009): 426–37.

Moore, Sarah A. "Garbage Matters: Concepts in New Geographies of Wastes." *Progress in Human Geography* 36.6 (2012): 780–99.

Morath, Susan J. "Regulating Food Waste." *Texas Environmental Law Journal* 48.2 (2018): 1–35.

Musulin, Kristin. "The 2016 Dive Awards for the Waste Industry." *Waste Dive*, November 29, 2016.

Nagle, Robin. *Picking Up: On the Streets and Behind the Trucks with the Sanitation Workers of New York City*. New York: Farrar, Strauss and Giroux, 2013.

Nestle, Marion. *Food Politics: How the Food Industry Influences Nutrition and Health*. Berkeley, CA: University of California Press, 2002.

Newell, Peter, and Adam Bumpus. "The Global Political Ecology of the Clean Development Mechanism." *Global Environmental Politics* 12.4 (2012): 49–67.

NRDC/Harvard Food Law and Policy Clinic. *The Dating Game: How Confusing Date Labels Lead to Food Waste in America*. NRDC Report. 2013.

O'Neill, Kate. "Regulations as Arbiters of Risk: Great Britain, Germany, and the Hazardous Waste Trade in Western Europe." *International Studies Quarterly* 41.4 (1997): 687–718.

O'Neill, Kate. *Waste Trading among Rich Nations: Building a New Theory of Environmental Regulation*. Cambridge, MA: MIT Press, 2000.

O'Neill, Kate. "Will China's Crackdown on 'Foreign Garbage' Force Wealthy Countries to Recycle More of Their Own Waste?" *The Conversation*, December 13, 2017a.

O'Neill, Kate. *The Environment and International Relations* (2nd edn). Cambridge: Cambridge University Press, 2017b.

O'Neill, Kate. "Explainer: The Plastic Waste Crisis Is an Opportunity for the US to Get Serious About Recycling at Home." *The Conversation*, August 17, 2018.

O'Neill, Kate. "Linking Wastes and Climate Change: Bandwagoning, Contention and Global Governance." *WIRES Climate Change* 10.2 (2019): 1–17.

Ottoviani, Jacopo. "E-Waste Republic." *AlJazeera.com*, 2015.

Pacheco-Vega, Raul, and Kate Parizeau. "Doubly Engaged Ethnography: Opportunities and Challenges When Working with Vulnerable Communities." *International Journal of Qualitative Methods* 17.1 (2018): 1–13.

Parizeau, Kate. "Urban Political Ecologies of Informal Recyclers' Health in Buenos Aires, Argentina." *Health & Place* (2015): 67–74.

Paterson, Matthew. "Where There's Muck There's Brass." *Cahiers de l'idiotie: Merde* 5 (2012): 77–87.

Pellow, David Naguib. *Garbage Wars: The Struggle for Environmental Justice in Chicago*. Cambridge, MA: MIT Press, 2002.

Peluso, Nancy Lee. "Plantations and Mines: Resource Frontiers and the Politics of the Smallholder Slot." *The Journal of Peasant Studies* 44.4 (2017): 834–69.

Porpino, Gustavo, Juracy Parente, and Brian Wansink. "Food Waste Paradox: Antecedents of Food Disposal in Low Income Households." *International Journal of Consumer Studies* 39 (2015): 619–29.

Postrel, Virginia, and Adam Minter. "The Future of Clothing Isn't in Tatters." *Bloomberg.com*, March 31, 2018.

Princen, Thomas. "Distancing: Consumption and the Severing of Feedback." In *Confronting Consumption*. Eds. Princen, Thomas, Michael F. Maniates and Ken Conca. Cambridge, MA: MIT Press, 2002, pp. 103–32.

Princen, Thomas, Michael F. Maniates, and Ken Conca, eds. *Confronting Consumption*. Cambridge, MA: MIT Press, 2002.

Puckett, Jim, Leslie Byster, Sarah Westervelt, et al. *Exporting Harm: The High-Tech Trashing of Asia*. Seattle, WA: Basel Action Network, 2002.

Puckett, Jim, Sarah Westervelt, Richard Gutierrez, and Yuka Takamiya. *The Digital Dump: Exporting Re-Use and Abuse to Africa*. Seattle, WA: Basel Action Network, 2005.

Rathje, William, and Cullen Murphy. *Rubbish! The Archaeology of Garbage*. Vol. Tucson, AZ: The University of Arizona Press, 2001.

Reagan, Robert. "A Comparison of E-Waste Extended Producer Responsibility Laws in the European Union and China." *Vermont Journal of International Law* 16 (2015): 662–86.

Reddy, Rajyashree N. "Reimagining E-Waste Circuits: Calculation, Mobile Policies, and the Move to Urban Mining in Global South Cities." *Urban Geography* 37.1 (2016): 57–76.

Reno, Joshua. *Waste Away: Working and Living with a North American Landfill*. Berkeley, CA: University of California Press, 2016.

Resnick, Elana. "Discarded Europe: Money, Trash and the Possibilities of a New Temporality." *Anthropological Journal of European Cultures* 24.1 (2015): 123–31.

Rogoff, Marc J., and Ian Spurlock. "Employment in a Tech Savvy Waste Industry." *MSW Management* (September/October 2017): 54–8.

Rootes, Christopher. "From Local Conflict to National Issue: When and How Environmental Campaigns Succeed in Transcending the Local." *Environmental Politics* 22.1 (2013): 95–114.

Rosengren, Cole. "What's Next for the NYC Commercial Waste World?" *Waste Dive*, August 19, 2016.

Rosengren, Cole. "Special Investigation: Living on Scraps." *DigBoston*, July 12, 2018a.

Rosengren, Cole. "Could New Tariffs Be the Final Blow for US–China Scrap Trade?" *Waste Dive*, August 9, 2018b.

Sánchez, Tania Espinoza. "Threat of a New Waste Incinerator in Mexico City Puts Informal Waste Pickers' Livelihoods at Risk." *WIEGO Blog*, March 1, 2018.

Saphores, Jean-Daniel M., Hilary Nixon, Oladele Ogunseitan, and Andrew A. Shapiro. "How Much E-Waste Is There in US Basements and Attics? Results from a National Survey." *Journal of Environmental Management* 90 (2009): 3322–31.

Saviano, Roberto. *Gomorrah: A Personal Journey into the Violent International Empire of Naples' Organized Crime System*. London: Picador, 2008.

Scheinberg, A., M. Simpson, Y. Gupt, et al. *Economic Aspects of the Informal Sector in Solid Waste Management*. Eschborn, Germany: GTZ and CWG, 2010.

Schenck, Catherina, Nik Theodore, Philip F. Blaauw, Elizabeth C. Swart, and Jacoba M.M. Viljoen. "The N2 Scrap Collectors: Assessing the Viability of Informal Recycling Using the Sustainable Livelihoods Framing." *Community Development Journal* 53.4 (2017): 656–74.

Schneider, Felicitas. "The Evolution of Food Donation with Respect to Waste Prevention." *Waste Management* 33 (2013): 755–63.

Schröder, Patrick, Manisha Anantharaman, Kartika Anggraeni, and Timothy J. Foxon, eds. *The Circular Economy and the Global South: Sustainable Livelihoods and Green Industrial Development.* London: Earthscan, 2019.

Schumpeter, Joseph A. *The Theory of Economic Development: An Inquiry into Profits, Capital, Credit, Interest, and the Business Cycle.* Oxford: Oxford University Press, 1961.

Schwabl, P., B. Liebmann, S. Köppel, et al. "Assessment of Microplastic Concentrations in Human Stool – Preliminary Results of a Prospective Study." *United European Gastroenterology Week 2018.*

Sheppard, Kate. "E-Waste Recycling in US Prisons." *Grist*, December 23, 2006.

Slackman, Michael. "Cleaning Cairo, but Taking a Livelihood," *The New York Times*, May 24, 2009.

Smithers, Rebecca. "Salad Days Soon Over: Consumers Throw Away 40% of Bagged Leaves." *The Guardian*, May 24, 2017.

Soma, Tammara. "Gifting, Ridding and the 'Everyday Mundane': The Role of Class and Privilege in Food Waste Generation in Indonesia." *Local Environment* 22.12 (2017): 1444–60.

Soma, Tammara, and Keith Lee. "From 'Farm-to-Table' to 'Farm to Dump': Emerging Research on Household Food Waste in the Global South." *Conversation in Food Studies.* Eds. Anderson, Colin R., Jennifer Brady and Charles Z. Levkoe. Winnipeg: University of Manitoba Press, 2016.

Spierling, Sebastian, Carolin Röttger, Venkateshwaran Venkatachalam, et al. "Bio-Based Plastics – A Building Block for the Circular Economy?" *Procedia CIRP* 69 (2018): 573–8.

Stahel, Walter R. "Circular Economy." *Nature* 531.7595 (2016) 435–8.

Stern, Rachel E. *Environmental Litigation in China: A Study in Political Ambivalence.* Cambridge: Cambridge University Press, 2013.

Strasser, Susan. *Waste and Want: A Social History of Trash.* New York: Metropolitan Books, 1999.

Sundt, Peter, Per-Erik Schulze, and Frode Syversen. *Sources of Microplastics Pollution to the Marine Environment.* Report produced for the Norwegian Environment Agency, 2014.

Szasz, Andrew. "Corporations, Organized Crime and the Disposal of Hazardous Waste: An Examination of the Marking of a

Criminogenic Regulatory Structure." *Criminology* 24.1 (1986): 1–27.

Toscano, Nick. "China's Import Ban Shakes Up Plastic Waste War." *The Age*, September 10, 2018.

Uddin, Sayed Mohammad Nazim, and Jutta Gutberlet. "Livelihoods and Health Status of Informal Recyclers in Mongolia." *Resources, Conservation & Recycling* 134 (2018): 1–9.

UN-Habitat. *The Challenge of Slums: Global Report on Human Settlements 2003*. London: Earthscan, 2003.

UNFAO. *Global Food Losses and Food Waste – Extent, Causes and Prevention*. Rome: UN Food and Agriculture Organization, 2011.

UNFAO. *Food Wastage Footprint: Impacts on Natural Resources: Summary Report*. Rome: UN Food and Agriculture Organization, 2013.

United Church of Christ Commission for Racial Justice. "Toxic Wastes and Race in the United States: A National Report on the Racial and Socioeconomic Characteristics of Communities with Hazardous Waste Sites." 1987.

UNODC. *Transnational Organized Crime in East Asia and the Pacific: A Threat Assessment*. Vienna: United Nations Office on Drugs and Crime, 2013.

US Environmental Protection Agency. "How Communities Have Defined Zero Waste". n.d. Available at https://www.epa.gov/transforming-waste-tool/how-communities-have-defined-zero-waste.

Vallette, Jim, and Heather Spalding. *The International Trade in Hazardous Wastes: A Greenpeace Inventory* (5th edn). Washington, DC: Greenpeace International Waste Trade Project, 1990.

Velis, Costas. *Global Recycling Markets: Plastic Waste. A Story for One Player – China*. Vienna: ISWA Taskforce on Globalization and Waste Management, 2014.

Vergara, Sintana E., and George Tchobanoglous. "Municipal Solid Waste and the Environment: A Global Perspective." *Annual Review of Environment and Resources* 17 (2012): 277–309.

Vilella, Mariel. *The European Union's Double Standards on Waste Management and Climate Policy: Why the EU Should Stop Buying CDM Carbon Credits from Incinerators and Landfills in the Global South*. Brussels: Global Alliance for Incinerator Alternatives, 2012.

Watts, Jonathan. "Eight Months on, Is the World's Most Drastic Plastic Bag Ban Working?" *The Guardian*, April 25, 2018.

Wilson, David C. "Development Drivers for Waste Management." *Waste Management & Research* 25 (2007): 198–207.

Wilson, David C., Adebisi O. Araba, Kaine Chinwah, and Christopher R. Cheeseman. "Building Recycling Rates through the Informal Sector." *Waste Management* 29.2 (2009): 629–35.

Wilson, David C., Ljiljana Rodic, Prasad Modak, et al. *Global Waste Management Outlook*. Vienna: United Nations Environment Programme/International Solid Waste Association, 2015.

Winslow, Kevin M., Steven J. Laux, and Timothy G. Townsend. "A Review on the Growing Concern and Potential Management Strategies of Waste Lithium-Ion Batteries." *Resources, Conservation & Recycling* 129 (2018): 263–77.

Winter, Debra. "The Violent Afterlife of a Recycled Plastic Bottle." *Atlantic Monthly*, December 4, 2015.

Wittmer, Josie, and Kate Parizeau. "Informal Recyclers' Geographies of Surviving Neoliberal Urbanism in Vancouver BC." *Applied Geography* 66 (2016): 92–9.

World Economic Forum, Ellen MacArthur Foundation, and McKinsey & Company. *The New Plastics Economy – Rethinking the Future of Plastics*. 2016. Available at http://www.ellenmacarthur foundation.org/publications

WRAP. "Wrap and the Circular Economy". n.d. Available at http://www.wrap.org.uk/about-us/about/wrap-and-circular-economy

Wynne, Brian, ed. *Risk Management and Hazardous Waste: Implementation and the Dialectics of Credibility*. Berlin: Springer-Verlag, 1987.

Wynne, Brian. "The Toxic Waste Trade: International Regulatory Issues and Options." *Third World Quarterly* 11.3 (1989): 120–46.

Xiarchos, Irine M., and Jerald J. Fletcher. "Price and Volatility Transmission between Primary and Scrap Metal Markets." *Resources, Conservation & Recycling* 53 (2009): 664–73.

Xue, Li, Gang Liu, Julian Parfitt, et al. "Missing Food, Missing Data? A Critical Review of Global Food Losses and Food Waste Data." *Environmental Science and Technology* 51 (2017): 6618–33.

Yee, Amy. "In Sweden, Trash Heats Homes, Powers Buses and Fuels Taxi Fleets." *The New York Times*, September 21, 2018.

Zalasiewicz, Jan, Mark Williams, Colin N. Waters, et al. "Scale and Diversity of the Physical Technosphere: A Geological Perspective." *The Anthropocene Review* 4.1 (2017): 9–22.

Zhang, Shengen, Yunji Ding, Bo Liu, and Chein-chi Chang. "Supply and Demand of Some Critical Metals and Present Status of Their Recycling in WEEE." *Waste Management* 65 (2017): 113–27.

Zimring, Carl A. *Clean and White: A History of Environmental Racism in the United States.* New York: NYU Press, 2015.

Zinda, John Aloyisus, Yifei Li, and John Chung-En Liu. "China's Summons for Environmental Sociology." *Current Sociology* 66.6 (2018): 867–85.

Index